Understanding Social Work Research

Understanding
Social Work Research

Hugh McLaughlin

First published 2007
Reprinted 2007

SAGE Publications Ltd
1 Oliver's Yard
55 City Road
London EC1Y 1SP

SAGE Publications Inc.
2455 Teller Road
Thousand Oaks, California 91320

SAGE Publications India Pvt Ltd
B1/I 1 Mohan Cooperative Industrial Area
Mathura Road, New Delhi 110 044
India

SAGE Publications Asia-Pacific Pte Ltd
33 Pekin Street #02-01
Far East Square
Singapore 048763

British Library Cataloguing in Publication data

A catalogue record for this book is available from the British Library

ISBN: 978-1-4129-0848-1 (hbk)
ISBN: 978-1-4129-0849-8 (pbk)

Library of Congress control number 2006903773

Typeset by C&M Digitals (P) Ltd., Chennai, India
Printed on paper from sustainable resources
Printed and bound in Great Britain by Athenaeum Press Ltd., Gateshead, Tyne & Wear

I would like to dedicate this book to my father, Joseph McLaughlin, who always encouraged me to challenge received wisdoms and to critically examine any 'truths', no matter whose 'truth' they were.

Hugh

Contents

Acknowledgements		ix
1	Why Research for Social Work?	1
2	The Research Business	16
3	The Philosophy of Social Research	25
4	Ethical Issues in Social Work Research	46
5	Evidence-based Practice – Panacea or Pretence?	72
6	Service Users and Research – the Next Frontier	88
7	Research and Anti-oppressive Practice	114
8	Interdisciplinary Contributions to Social Work and Social Work Research – the New Orthodoxy	134
9	Getting Research into Practice	150
10	Whither Social Work Research – Challenges for the Third Millennium	177
Appendix: Abbreviations used in the Book		187
References		189
Index		205

Acknowledgements

In writing this book I would like to thank all my ex-colleagues from my days in practice and current colleagues in academia who have influenced me more than they can know and helped to sustain my interest and passion in research for social work. I would also like to thank Rachel Burrows, Zoe Elliott and Anna Luker from Sage for their editorial encouragement and prompt response to any query. In particular I would like to thank Catherine for her love, support and willingness to undertake other tasks whilst yet again I disappeared to the study, and to James and Eleanor who helped keep my feet on the ground and for reminding me that being a taxi-driver for one's children is just as important as being an academic.

1 Why Research for Social Work?

This book seeks to identify research as an underused but essential tool for the busy social worker in undertaking their difficult, demanding and often contradictory tasks within society. For too long research has been ignored by social workers or at best treated as an add-on or luxury. There are many reasons why this has been so and some of these will be explored in later chapters. Social workers work with some of the most vulnerable people within society – children who have been abused, older people with Alzheimer's disease or those suffering chronic mental ill health. All of these vulnerable people deserve the highest professional expertise from their social workers and this book argues that this is not possible without also including a research-minded perspective. Research can help social workers deliver their practice agenda and in many ways good research and good social work are synonymous. D'Cruz and Jones (2004: 2) write:

> We teach research to social work students because we believe that social work practice is more likely to be effective when social workers are able to draw on and evaluate previous research.

As a social worker, or social work student, you will want to know whether your interventions are making a difference, either positively or negatively. Social work is not a neutral activity. Social work can lead to a positive outcome for service user or social worker or it may lead to a negative outcome for one or both of them. To this end it is imperative for the social worker to be in a continuous reflective relationship with their practice seeking to find evidence and answers that help them to identify whether their intervention is effective or merely interference. As such the social worker may have to ask:

- Has my task-centred intervention helped safeguard a child previously abused?
- Will the care package I commission enable Mr and Mrs Jones to live independently in their own home?
- Is the Council's policy on drug misuse working?
- Does respite care work?
- What does the research evidence say about effective residential provision?

As a social worker there is a need to be able to ask these and similar questions at both the individual – your own caseload, the agency and the policy level. The policy level may be at local service delivery or maybe at a national level, for example, your authority's position on achieving the government's performance target of doubling the number of children adopted or the impact of implementing *Fair Access to Care Services* (Department of Health, 2001a).

The Notion of the Social Worker and Social Work

The notion of a social worker has been defined in a number of different ways over the years. In 1982 following a series of child abuse tragedies and inquiries the then UK Conservative Government commissioned Peter Barclay to review the role and function of social workers. In the prelude to the report Barclay noted:

> Too much is expected of social workers. We load them with unrealistic expectations and then complain when they do not live up to them. Social work is a relatively young profession. It has grown rapidly as the flow of legislation has greatly increased the range and complexity of its work. (Barclay, 1982: vii)

Payne (1997) describes three major differing approaches to how we conceptualise social work. These are:

- Individualism–reformism: this refers to social work as an activity that aims to meet social welfare needs on an individualised basis.
- Socialist collectives: this approach focuses on promoting co-operation within society in order that the most oppressed and disadvantaged can gain power and take control of their own lives.
- Reflexive–therapeutic: this approach is focused on promoting and facilitating personal growth so that people are enabled to deal with the suffering and disadvantage they experience.

It is not the intention of this book to support one definition at the expense of the others but to show that social work is contested and can be conceptualised in a range of different ways. More recently the International Association of Schools of Social Work and the International Federation of Social Workers agreed the following definition as they sought to transcend national boundaries and identify social work within a global economy:

> The social work profession promotes social change, problem solving in human relationships, and the empowerment and liberation of people to enhance

> well-being. Utilising theories of human behaviour and social systems, social work intervenes at points where people interact with their environments. Principles of human rights and social justice are fundamental to social work. (IASSW and IFSW, 2004: 4)

Whilst this is the agreed definition by the different international social work associations, their members do accept that how it will be interpreted will vary between nations and contexts. A respondent to the consultation from Hong Kong proposed additions including an acknowledgement of responsibility and the importance of the collective.

An alternative view of the definition of the social work task comes from Thompson in his popular book *Understanding Social Work* where he argues that what constitutes social work is 'up for grabs'. In particular he emphasises the extent to which powerful groups and institutions define what constitutes social work and define the limits of policy and practice. In answer to this he seeks to identify key themes related to existentialism. In particular he argues that social work should be:

- ontological – sensitive to the personal and social dimensions and the interactions between the two;
- problem-focused – sensitive and responsive to the existential challenges we all face, but particularly those that are related to social location and social divisions;
- systematic – with a clear focus on what we are doing and why (our goals and our plans for achieving them);
- reflective – open minded, carefully thought through and a source of constant learning rather than a rigid, routinised approach to practice;
- emancipatory – attuned to issues of inequality, discrimination and oppression, and geared towards countering them where possible. (Thompson, 2000: 23)

Both these definitions accept that they are not written on tablets of stone and are likely to change as society changes. This is not to suggest an 'anything goes' mentality but to be realistic that changes in society will result in changes in social work practice. Social work is not a fixed entity. Therefore the view of what social work is, or should be, becomes revised and redrawn. The difficulty remains at what stage do the changes become fundamental and social work no longer exists? Or, to put it more concisely, what do you consider to be the core of social work? You might like to write this down now and keep it somewhere safe to look at again as you finish this book or even a couple of years into the future to see if it has changed.

The former of the two definitions, the IASSW and IFSW (2004), may be criticised for its failure to identify the 'control' elements of social work, in particular the failure to identify the statutory powers available to social workers in intervening in situations with vulnerable children and young people, to enforce mental health treatment and services or to protect vulnerable older people (Horner, 2003). Thompson's definition can also be criticised as it is unclear how each of his themes interact and in particular if they should conflict as to which would have precedence. The purpose of identifying these two authorities and their differing approaches to the definition of social work is not to say that one is right and one is wrong, but to begin to identify the complicated reality of social work, which by its very nature exists in a world of complex interactions, contradictions and ambiguities. Butrym (1976), 30 years ago recognised that the nature of social work did not exist separately from what social workers did – practice shapes the profession's identity. This is not to say that social work is equated to what social workers do but that there is a strong relationship. Social work is more than what social workers do. Yet the very nature of this practice is shaped by the society in which the practitioner exists and the interaction between the two.

> The preoccupation of social work with people and their social circumstances creates its main occupational risk – a lack of specificity, an inherent ambiguity – which makes it particularly prone to changes and fluctuations, not all of which are necessarily consistent or logical. (Butrym, 1976: ix)

What is important for our future discussions is that social work is neither a neutral nor static activity, it can have positive and/or negative consequences and represents a values-driven form of welfare practice.

Social Work and the Modernisation Agenda

Within the British context social service departments were created following the introduction of the Local Authority Social Services Act 1970. This brought together the three separate local council departments: children, mental health and welfare into a single unitary social work department headed up by a Director of Social Services. Each local authority had to have a designated Director of Social Services by statute. Initially, social work was viewed as a generic activity but has become increasingly specialist and fragmented ever since with the current situation of the break up of social service departments and recombination into differing organisational structures.

As previously noted, social work has experienced increasing demands for change. Major structural changes occurred in the early 1990s signalled by the introduction of the Children Act 1989 and the NHS and Community Care Act 1990. These two acts pushed many authorities into separating children and adult services. The Children Act 1989 was one of the first major pieces of social services legislation to be explicitly informed by research (Parker et al., 1991). The introduction of these two acts increased the fragmentation of the service and subjected it (albeit unevenly) to the disciplines of the mixed economy of welfare. This involved redefining the role of social services from that of provider of services to that of 'enabler', 'commissioner' or 'purchaser' or services. The government also sought to promote a mixed economy of care whereby the 'commissioners' and 'purchasers' of service were expected to commission or purchase services less from the local authority and more from a mix of public, private and voluntary providers. There was also a broad intention to transform the nature of social services from a welfare agency run by professionals, allegedly too much in their own interests, to a customer-centred organisation run by professional managers. Adams sums up this trend as 'new managerialism' whereby:

> 'Managerialism' is the term used to describe styles of management which put managers in the central role of the organisation. 'New managerialism' (Adams, 1998: 44) is the term used to describe management approaches in the public services which were imported from the private sector in the early 1980s, with decisions based on the criteria of economy, efficiency and effectiveness. (Adams, 2002: 176)

Following the defeat of the Thatcher and Major Conservative governments, the Labour Party came to power in 1997 and far from retreating on the reformation of local government have moved it onto a new level. The Labour Government's big idea can be summed up in the discourse of modernisation. Modernisation has permeated all aspects of central and local government including social services and health services. Modernisation is represented in a range of publications (for example, DETR, 1998; Cabinet Office, 1990; Department of Health, 1998) as the necessary process for updating services to match the expectations of modern day consumers. It continues the attack on provider dominance, seeks to sharpen accountability and maintains the focus on business solutions to social and policy problems (Newman, 2000). Modernisation emphasises partnership (Balloch and Taylor, 2001; Glendinning et al., 2002) and joined-up thinking (Frost, 2005) whereby traditional boundaries which have been seen as barriers to meeting service users' needs are now addressed by workers working together in partnerships or joint teams. This emphasis upon partnerships and joined-up thinking accepted that traditional uni-professional ways of working were not going to address society's 'wicked' problems like poverty and social exclusion. Examples of this way of working can be found in mental health trusts, learning disability trusts and youth offending teams (YOT) all of which contain a range of professional workers working together on the same problems. In the case of youth offending teams this may consist of social workers, probation officers, police officers, educational representatives and health representatives. YOTs were located under the leadership of local authority chief executives to emphasise their multiprofessional nature as opposed to being located with any of the professional groups from whom its membership originates from.

Denise Platt, when Chief Inspector for Social Services, noted in her annual report:

> The present government is committed to reforming public services. Its vision is of public services where the services are designed around the needs of the people who use them, rooted in the values of the community. To deliver the agenda and to modernise the service we are asking the people who work in social services to work in new ways. (Platt, 2002: para 1.4–1.5)

In order to promote the Government's agenda in England a star rating system was introduced in 2000 bringing together a range of performance data so that the public and service users can see at a glance a simple summary of performance. The star rating of councils includes the performance of social service departments who are also rated separately. Social service department ratings include two major means of quality control. First, there is the programme of inspections undertaken by the Commission for Social Care Inspections (CSCI) alone, the Joint Area Reviews (JARs) undertaken by CSCI and nine other inspectorates. Secondly, there is the Performance Assessment Framework (PAF) that covers key indicators for all aspects of social service performance. From an amalgamation of both of these sources, cross-referenced with the Best Value performance indicators, a social services department's star rating is given. Best Value is the process by which local authorities are expected to critically review their services to ensure they are effective and efficient and are provided economically. Economic need not be interpreted as the cheapest price if value for money can be demonstrated.

Alongside this growth in performance indicators there has been increasing government support for practice based on research evidence. Taking the example of child and family services this has been particularly visible since the Children Act (1989). Besides individually funding research projects the Department of Health has also developed a number of research overviews summarised for practitioners and managers including: *Patterns and Outcomes in Child Placements* (1991), *Child Protection: Messages from Research* (1995) and *Caring for Children Away from Home: Messages from Research* (1998). In recent years this has also spread into other research dissemination activities including conferences, the support of Research in Practice (RiP), Making Research Count (MRC), the launch of the Social Care Institute for Excellence (SCIE), the development of the Scottish Institute for Excellence in Social Work Education (SIESWE) and the production of research briefings and websites providing advice and information for social work practitioners (for example, www.dfes.gov.uk/choiceprotects). Butler has wryly commented that:

> It is rare to complete a student placement visit to a social services department without at least coming across one of the several Department of Health (DoH) 'Messages from Research' (e.g. DoH, 2001b) or 'Quality Protects' (e.g. DoH/RIP/MRC, 2002) briefings that seek to influence social work practice with children and families through the use of research findings. Central government is actively involved in other ways too: not just in funding and disseminating research but also in various forms of its governance and ordering. It is difficult to remember a time when interest in social work research was so widespread, so urgent and so apparently full of possibilities. (Butler, 2003: 19)

What Butler doesn't tell us is whether these documents were being actively read or whether they were merely available and gathering dust on bookshelves or desks. Alongside these initiatives there has been the launch of SCIE, an independent company and charity, funded partly by the government. SCIE is committed to developing and promoting knowledge about good practice in social care. SCIE is very much a part of the government's drive to modernise social care and is currently commissioning and publishing a series of knowledge reviews and reports about how knowledge can be used to support social care. In Scotland the nine universities providing social work training have come together to establish the SIESWE whose remit is to promote a strategic approach to developing and ensuring quality in the learning and knowledge transfer in the implementation of the social work degree in Scotland (www.sieswe.org.uk). The SCIE reviews can be downloaded free from the World Wide Web (www.scie.org.uk) or may be obtained free from SCIE direct. Knowledge Review number 3 identifies the *Types and Quality of Knowledge in Social Care* (Pawson et al., 2003). Pawson and colleagues identified a classificatory framework for understanding knowledge in social work and social care. In particular their model identified:

- Organisational knowledge: knowledge gained from the governance and regulation activities involved in organising social care.
- Practitioner knowledge: knowledge gained by practitioners in their day-to-day work which tends to be personal, tacit and context specific.
- User knowledge: knowledge gained from the experience of using social care services; again this is often tacit and is explored further in **Chapter 6**.
- Research knowledge: knowledge gathered systematically within a planned strategy; such knowledge is mostly explicit and available in reports, evaluations, books and articles.
- Policy community knowledge: knowledge that is gained from the wider policy context and may include knowledge in the civil service, think tanks and agencies.

This book is primarily interested in the use of research knowledge, but as the reader will realise, although Pawson et al. (2003) have identified a conceptually distinct framework, there are often overlaps. In particular we will look at practitioner knowledge and service user knowledge, but will also be aware of the potential for organisational knowledge and policy community knowledge to impact upon social work research.

At this point it is worthwhile clarifying the distinction between the social care sector and social work. In this book social workers refer to those with a professional qualification in social work and are registered with their local regulatory body. Whilst this group are also part of the wider social care workforce, they represent only a small part, as this is primarily made up of unqualified staff many of whom work part-time. The *Social Services Workforce Survey* (Social Care and Health Workforce Group, 2004) in 2003 identified a total of 276,490 staff employed by social services departments in England out of a social care workforce of approximately 1.4 million. The vast majority of local authority staff were female (80.7 per cent) and white (90 per cent). When the report concentrated on field social workers, occupational therapists, homecare staff, managers and care staff in residential setting they identified a total of 150,185 employees of whom only 44,155 were qualified and just over half of whom were employed in childcare. From these statistics it is very clear that the vast majority of staff are not qualified in social work. Social care workers may include homecarers, residential workers, support workers, social service officers and family centre workers who may, or may not, have a qualification, but it is only those who are employed in posts which require a social work qualification that we refer to as social workers in this book.

In the UK the General Social Care Council (England), the Care Council for Wales, the Scottish Social Services Council and the Northern Ireland Social Care Council have been tasked to register and regulate the activities of all social care workers beginning with qualified and student social workers. Also from April 2005 it was illegal to use the title of social worker unless someone has an approved social work qualification and is registered with the appropriate regulatory body. In many other countries like America, Australia and Germany social workers already have protection of title and a governing registration body.

Social work in the UK is presently at a point of contradiction. The Labour Government has brought in a new graduate requirement for qualifying social workers, has protected the title of social worker, established registration of social workers and a requirement for continuing professional development. At the same time we are experiencing the fragmentation and dismantling of stand-alone social service departments, previously the main employment opportunity for social workers. These stand-alone social service departments have been replaced by new local authority configurations of education and children's social services representing a split from adult social services. In many of these authorities the children's services have been headed up by Directors of Education with the Directors of Social Services heading up the adult services, which in some cases has also included housing or other community responsibilities. It will be interesting to observe and evaluate how these changing structural arrangements will impact upon the social work profession. These arrangements have been further complicated where social service personnel have been merged with health personnel to form multidisciplinary teams in mental health and learning disability trusts. In some parts of the country children's social services staff along with education department personnel and health colleagues are being combined to form Children's Trusts. Today, social workers are employed in an increasingly wide range of non-statutory social care agencies. These include those in the private and voluntary sector, multi-agency teams, Sure Start, Connexions, Youth Offending Teams, Drug and Alcohol Teams, Mental Health Trusts, Children and Family Court Advice and Support Services (CAFCASS) and an ever growing number of recruitment agencies etc. For many of these organisations social work will not be the major profession nor social care the dominant discipline. The majority of social workers in the future will not be employed in monolithic social service departments but in a mosaic of welfare provision. This changing terrain of social work practice makes it even more important that social workers are able to use research effectively and identify the distinctive nature of social work knowledge and practice or risk becoming subsumed by those professions who can.

The above relates particularly to England, but England is only one part of the four nations that make up the UK who all organise things slightly differently. In Scotland the term 'social work services' is used to describe the full range of services including criminal justice as well as child and adult services. Whereas criminal justice is separate in the rest of the UK, Northern Ireland's social services have developed against a backdrop of a contested nation and civil unrest. 'The Troubles' and political uncertainty have resulted in social services being located under the direction of the Department of Health and Social Services and managed as an integrated service with health by Health and Social Services Boards since 1973. Wales, since devolution, has seen a great deal of policy activity and a divergence from the English context including a decision not to promote the development of Children's Trusts. A particular issue within the Welsh context has been the promotion of the Welsh language and the need to promote Welsh-speaking services for those whose first language is Welsh. (For a fuller discussion on social work in the British Isles see Payne and Shardlow, 2002.)

This has been a rather long introduction into the state of social work in the UK, but provides the context in which social work research is expected to operate and promote research minded practice.

Reflexive Questions

You might like to consider what type of setting you work in or would ideally wish to work in. Why have you chosen this type of setting?

If you have not chosen the 'traditional' local authority department, why have you done so?

If you have chosen local authority social work what do you see as the advantages of such a system?

What is Research?

Up until now it has been assumed that we all have a shared understanding about what research is. The stereotypical image of the researcher emphasises what one might call the manipulative aspects of the role designing questionnaires or engaging in statistical analysis, writing incomprehensible reports and detached from reality living in an ivory spire. At best, social work practitioners may experience research as irrelevant and inaccessible and at worst as the process of being manipulated for the researcher's career enhancement.

> **Reflexive Questions**
>
> What is your image of the researcher?
>
> How do you define research?
>
> Who undertakes research?
>
> Which organisations undertake research?

As Hughes (1990) notes, research is carried out to discover something about the world and is an activity that many of us engage in daily as we seek to find out something we did not know previously. What makes research different from other ways of finding out is its processes that are characterised as systematic and disciplined. As Becker and Bryman (2004: 14) observe for something to count as research:

> the enquiry must be done in a systematic, disciplined and rigorous way, making use of the most appropriate research methods and designs to answer specific research questions.

The Research Governance Framework for Health and Social Care's (RGF), definition is similarly technical in its application:

> Research can be defined as the attempt to derive generisable new knowledge by addressing clearly defined questions with systematic and rigorous methods. (Department of Health, 2001b: para 1.7)

These technical definitions are at some variance with our view of social work that highlights its value-driven nature. At one level it could be argued that this does not matter and that research should be seen as standing objectively outside the practice milieu. To do so would be to privilege the researcher's perspective of the world. Research like social work is value-driven. As Rosen has noted:

> The selection of a research topic and corresponding method are in many ways a life choice. They are indicative of what the researcher believes as important to 'see' in the world, to investigate and to know. An individual choice of topic and method corresponds to his or her ontological vision, his or her model of being. (Rosen, 1991: 21)

The question of which topic to research is not purely a technical question, but also a philosophical one. In deciding upon a topic to research and the resultant research methodology the researcher is making a statement about human nature, society and the relationship between the two. Likewise we need to acknowledge that research sponsors and funders will target funding on particular aspects of social life and organisations and will only allow access to particular aspects of their organisations. It is more likely to be easier to gain access to research how social workers could become more efficient and effective as opposed to similarly examining the impact of senior managers, councillors or management boards.

Coupled to this there is also the vested interest in organisations to ensure research reports reflect a positive view of the organisation. Recently I was 'invited' to meet with an Assistant Director immediately prior to a research report being agreed for publication. The local authority did not like what the report said about their management team or style and felt the report was too sympathetic and accepting of the views of their front-line staff. The local authority wanted the final pages of the report rewritten before they would agree to their organisation's data being included in the report. Following long and tortuous negotiations a compromise was agreed which allowed both parties to retain their integrity. This can become even more invidious when research commissioners block publication because they do not want the results of the research to be in the public arena. Such checks and balances can be helpful in ensuring accuracy in research, but at the other extreme they can also result in compromising and undermining the veracity of the research process and any contributions to knowledge the research may make. Researchers and organisational commissioners need to clearly identify the parameters around publication at the beginning of the research process. These need to ensure that research publications do not become blocked because the results are not positive whilst researchers also need to be able to convince organisations of their political awareness, accuracy and honesty of their work. This topic is dealt with in greater detail in **Chapter 9**.

Research Mindedness

In order to address these difficulties Everitt et al. (1992) identified 'research mindedness' as a means of enhancing good social work practices. They also suggested that social work and social work research could be conceptualised as interrelated processes. In order to achieve this they identified three principles of 'research mindedness':

- A participatory/developmental rather than a social control model of social work.
- Anti-oppressive values are applied.
- Genuine partnerships are established with those who social work serves.

These principles are in keeping with our previous discussions on the ethical nature of social work. It also emphasises that research subjects are valued in their own right and not purely viewed as objects to be exploited for the further advancement of the researcher. This view also suggests that the basis of 'knowing' is shared and should be made explicit between researcher, informant and the research audience. They also argue that subjectivity viewed as personal experience, or worldview, is valued and is seen as integral to the research experience. Everitt et al. (1992) did not make a claim for a definitive methodology but rather some tentative approximations and provisional suggestions. This book, like D'Cruz and Jones (2004), adds to the development of their research-minded practitioner who is seen as essential for the development and long-term survival of social work within these fragmented and contested times.

Social Work Research – A Different Type of Research?

Having identified the nature of social work, research and research mindedness, it is important to question whether there is a distinct form of research that can be termed social work research.

Reflexive Questions

What do you think?

Is there a distinct form of research that can be termed social work research?

If so, how would you identify whether a research project constituted social work research or just social research?

It is clear that we have social work academics that conduct social research into social work problems. It is also clear that others, who are not social workers, also research many of these issues, like domestic violence, child protection or disability. It is thus worth asking what makes the difference between a social work researcher and a non-social work researcher? Dominelli (2005) argues that there is a distinctive research approach that can be defined as social work research. In particular she identifies the following key features:

- A change orientation.
- A more egalitarian relationship between themselves and those who are the objects of their research.
- Accountability to 'clients'/service users for the products of their work.
- A holistic engagement with the different aspects of the problem(s) of people they are investigating. (Dominelli, 2005: 230)

None of these exclusively characterises social work research, but combined can be seen to typify its nature and scope. Social work research thus aims not only to support practice but also to transform it. Social work with its ethical imperative towards challenging social injustice requires social work researchers to examine both the socially excluded and the powerful elites who decide upon their social exclusion.

Social work research is also likely to include an emotional content that must also be taken into account. This will include adopting a holistic empowering approach to research subjects who are related to as active participants, not objects. Social work researchers are also keen to look at ways of promoting democracy and participation by involving service users in research design, delivery and dissemination (see **Chapter 6**).

A researcher researching female drug use would be interested in finding out the different types of drugs used, the prevalence of drug use, where and when drugs were used, why drugs were used and ways in which the dangers of drug use could be minimised or eliminated. A social work researcher researching the same subject would not only be interested in the same issues. A social work researcher would also consider how social work practice could change to become more effective and has to be ready to confront the ethical challenge of reporting a mother whose drug use was resulting in placing her children at risk of significant harm.

You may want to consider whether you feel Dominelli makes a compelling case for a distinctive social work research or not. It could be argued that nurses or lawyers researching drug abuse would be faced with the same dilemmas. This is only partly true; nurses and lawyers are not social workers and are more likely to view matters through a legal or health lens. They cannot be expected to understand social work processes to the same degree or to be able to make recommendations for changes in social work practice, except in a broad and general way. Whether there is, or not, something that can be called social work research is debated further in future chapters.

The future of social work is both uncertain and exciting; the previous certainties have been stripped away whilst the future remains in the balance. It has been argued that social work and social work research are value-driven activities, which are both contested and contestable. This is why it is so important for you as a social worker to become research-minded if you wish to be able to engage in debates about the effectiveness of your practice, to demonstrate the impact of government policies or to advocate on behalf of some of the most disadvantaged groups within our society. As a social worker it will not be enough to say that you believe something to be true, you need to be able to demonstrate the evidence for your belief whether this is in a contested childcare case or an accommodation review for a service user with a physical disability.

The Organisation of the Book

The book seeks to place social work research within a framework for research-minded practitioners who are able to see the challenges and opportunities provided by a better understanding of research and its potential. The book seeks to ground social work research within practice, its context, concepts and key issues. In order to maximise the benefit of each of the following chapters readers should regularly reflect on the following key questions.

Reflexive Questions

How does this help my understanding of social work and research?

What are the implications of this for my practice?

How might I become more research-minded?

How do we develop a research-mindedness within our team and/or service?

Throughout the book you will see reflexive questions like those above. These questions are there to help you to consider, reflect and develop your own views on some of the key issues facing social workers in coming to terms with research and research evidence. You may also want to consider writing your answers down to the different questions on sheets of A4 or a notebook to keep with this book to act as a reminder of your original thoughts.

Chapter 2 begins by locating social work as a practice discipline and examines the business of social work research. This lays the foundation for Chapter 3 concerning 'The Philosophy of Social Research' and in particular its major paradigms and how these impact upon research design and the research question. Chapter 4 then returns to the 'Ethical Issues in Social Work Research' part of which we have begun to discuss in this chapter. This chapter explores whether social work research can be seen to have a specific research code and identify the major ethical issues before, during and after a research project.

The book then moves onto identifying key issues in social work research beginning in Chapter 5 to examine the nature of evidence-based practice. Evidence-based practice is defined and explored from a medical model perspective before assessing its applicability to probation and social work. The politicisation of evidence-based social work is investigated along with the claims for a hierarchical evidence base. The next chapter investigates the trend towards involving service users in research identifying some of the practical and ethical problems for doing this whilst also seeking to identify when this would be appropriate and when it would not. The chapter also begins to identify when such an approach should be used and begins to examine what are its costs and benefits.

Chapter 7 focuses on anti-oppressive practice and research highlighting research considerations in relation to race and gender as exemplars of other sites of oppression. The chapter then seeks to develop an anti-oppressive research practice. Chapter 8 identifies the growing importance of interprofessional practice and interprofessional research especially as this is now an integral part of the requirements for the new social work award. In the next chapter there is a consideration of the growing importance given to 'Getting Research into Practice' and the barriers and drivers in relation to this. The final chapter provides a summary of the previous chapters and then examines the future for social work research – 'Whither Social Work Research: Challenges for the Third Millennium'.

Suggested Reading

D'Cruz, H. and Jones, M. (2004) *Social Work Research: Ethical and Political Contexts*. London: Sage. A very readable introduction grounded in the belief of improving social work practice through the application of research.

Thompson, N. (2000) *Understanding Social Work*. London: Macmillan. A widely available and easily readable introduction to the nature of social work practice.

2 The Research Business

This chapter begins by considering how research is funded and the process by which research becomes published. This then leads into a discussion on the nature of social work as a discipline, the contributory knowledge streams to social work and the requirements for social knowledge in the social work degree and Post-Qualifying (PQ) framework.

Why Bother with Research?

> **Reflexive Questions**
>
> Before going any further you might want to write down the reasons why you believe research is undertaken.
>
> Also you might like to consider who benefits from research?

On the surface this may appear a rather trivial question, but the answer is very important as it helps us understand what we expect of research, and in particular what we expect social work research to deliver. Your list may have included some of the following reasons:

- To answer questions that you did not know the answer to.
- To find out 'what works'.
- To inform decision making.
- To justify a particular intervention.
- To further the careers of researchers.
- To evaluate whether new interventions work.
- To make money.
- To win a Nobel Prize.
- For intellectual stimulation.

- To prove something doesn't work.
- To obtain the evidence to justify a change.
- For the completion of a dissertation.
- To enhance the reputation of social work as a serious discipline.
- For submission to the Research Assessment Exercise (RAE).

As can be seen from the above there are a wide variety of reasons why research gets undertaken. Some of these reasons focus around the intellectual desire to obtain knowledge in terms of answering the 'questions we do not know the answer to' or 'to evaluate new interventions'. There are also the reasons concerned with evaluating whether an intervention is working or not working. It is not unusual for a sponsor of research to commission research to obtain external evidence that something is working so that they can use the results in their marketing literature. This could be a case for an independent therapeutic children's residential home or within a new project supporting people with learning disabilities to live in the community. This has become all the more prevalent with the increasing auditing of services and the emphasis placed upon 'best value'. It may also happen that research is commissioned to prove something is not working to provide an external impetus for change. Managers and workers may be more willing to hear such messages about the shortcomings of their service than they would be if their own senior managers delivered the messages (Easterby-Smith et al., 1991).

In our list of reasons there were also the reasons concerned with furthering or developing the researcher's career or enhancing social work as a research discipline including submissions to the Research Assessment Exercise (RAE), of which more will be said later, or, even winning a Nobel Prize! There is also the pragmatic response in terms of completion of a dissertation or module. It should be noted that the choice of one reason does not necessarily exclude others so that you may be undertaking research to obtain an academic award but may also be genuinely interested in evaluating whether a new intervention works or not. Or, as a researcher you may be wanting to answer a question you do not know the answer to whilst also seeking to produce an internationally peer-reviewed article to help promote your career.

In terms of social work the instrumental reasons above take on a moral dimension when we remember that social work service users are amongst the most vulnerable in society. For the social worker, either using research or acting as a practitioner–researcher, there are also ethical dimensions to these questions (this will be discussed more fully in **Chapter 4**). It was noted in **Chapter 1** that a distinctive feature of social work was its value-driven nature and that social work research needed to reflect this. This is not to say that any research that does not reflect these values will not be of use to social workers. Patently, such a position would be untenable. As we have already shown social work is a complex activity with a set of contested meanings. It does though suggest that a social work researcher will adopt a particular stance in relation to the world, to the methods they will use and as to how service users and other research respondents will be treated in the research process.

The Research Business

Social work research is funded in a number of different ways. These include the funding universities receive via the Higher Education Funding Council (HEFC), competitive bidding to the research councils, government departments, charities or local social service providers. Researchers may also be directly approached by organisations to undertake research and lastly, there is unfunded research. This latter case may come about because of the particular interest of a researcher or for the completion of a higher degree or post-qualifying course. Unfunded research probably reflects the majority of practitioner research.

Not all research is undertaken in universities or by university staff. There are a growing number of independent research agencies and also some local authorities and voluntary agencies that have their own research staff. However, at present, the majority of research is undertaken within a university framework.

All higher education institutions receive annual funding from the Higher Education Funding Council (HEFC) that is primarily split into funding for teaching and learning and funding for research. In 2006–7 HEFC provided a total of approximately £1.3 billion for research funding as opposed to approximately £4.1 billion for teaching in England. In all but four of the 86 universities supported in this way the funding for teaching and learning was greater than that for research but in the Universities of Oxford, Cambridge, Imperial College of Science, Technology and Medicine and the University College London the positions were reversed with more funding for research than teaching and learning (*The Times* Higher Education Supplement, 2006). The research total for each centre is influenced by the outcome of the regular Research Assessment Exercises (RAE). The total research funding received by each university is given to the university to disperse in the way that it feels is most appropriate between its differing research centres and groups. This means a subject like social work will be competing for university research funding against other discipline areas including medicine, nuclear physics or politics.

The second area of funding identified was the research council. There are currently eight research councils of which the Economics and Social Research Council (ESRC) is the most important for social work. The ESRC is funded by government and aims to contribute to the economic competitiveness, the effectiveness of the public sector and quality of life in the UK (ESRC, 2006). The ESRC allocated

£77 million in 2004–5 via thematic priorities including social stability and exclusion, lifecourse, lifestyle and health and governance, and citizenship – all of which are of interest to social workers. Peer reviewers assess all applications for council grants. Shaw et al. (2004) undertook a review of the ESRC funding of social work and social care and found that researchers did not consider social work research, with its applied nature, was very attractive to the ESRC. To overcome some of the difficulties faced by social work it was recommended that social work should seek representation on the Research Grants Board, more social work researchers should become grant reviewers and that there needed to be pressure put on the ESRC to integrate the social work research agenda into programme development.

Thirdly, research may be commissioned by government by competitive application, as was the case in relation to research on the outcomes of the qualifying social work degree or the Sure Start programme. There is then research funded by charities including the Joseph Rowntree Foundation, Nuffield Foundation and Leverhulme Trust Foundation which all support research in social work. These foundations often have thematic approaches with competitive bidding alongside researcher generated research proposals.

Fourthly, research is also commissioned by local authorities and charities to focus on a specific issue. Often these relate to service evaluations, for example, an evaluation of the care planning approach in mental health or the effectiveness of direct payments to people with disabilities. These agencies may advertise the research contract or contact specific preferred researchers or research centres.

Lastly, the researcher may provide their own funding to research an issue that is of particular interest to them. Such research is often related to the attainment of a higher degree, but is not purely restricted to this. Practitioners and academics often undertake such research. Practitioners will probably focus on areas of particular interest possibly supported by their employers, but sometimes not. This support may include some or all of the following: paying for a higher degree, allowing the staff member time to undertake the research, rearranging their work schedule or providing access to research subjects within their organisation. Few authorities now have their own research staff and even where this is the case they will often be focused on areas to improve the organisation's performance indicators and/or star ratings.

Social work academics, unlike academics in the pure sciences, often do not have a PhD at their point of recruitment. Social work academics, in my experience, are primarily recruited from practice where a PhD or research experience is not a requirement and may in some cases be seen as a handicap. This results in social work academics being valued as much for their practice and teaching expertise as for their research ability. In many universities, potential social work lecturers do not have to be able to show a proven research track record only the potential to become research active. This has particular consequences for social work research as it seeks to establish a credible research culture and knowledge base whilst also competing with the more traditional academic departments in universities and higher education institutions. Mills et al.'s (2006) demographic review of the UK social sciences noted that social science academics tended to be older than their natural or physical sciences counterparts, that social work had the highest percentage (47 per cent) of staff aged between 50–55 and was one of seven areas highlighted as in need of greatest attention for developing capacity.

It is also worth considering the funding position between health and social care which are often located in the same faculty in many universities. Marsh and Fisher (2005: ix) graphically illustrate this when they note that:

> The overall spend per workforce member stands at about £25 in social care compared with £3,400 in health. Using the more specific comparison with primary care, the annual R&D spend per social worker is about £60, compared with £1,466 per general practitioner (GP); annual university research income from the Higher Education Funding Council (HEFC) Quality Related research (QR) is £8,650 per social work researcher and £26,343 per primary care researcher.

The UK Government has established an aspiration for the knowledge economy that 2.5 per cent of gross domestic product should be invested in research and development by 2014. This target is already met by health spending but would require an eight-fold increase in funding for social work and social care. These disparities, Marsh and Fisher (2005: ix) suggest, reduce the potential for social work research to deliver social welfare and 'hinders the modernisation of social care'.

This section has sought to show that the funding for research, limited as it is, comes from a diverse set of directions. It is also suggested that although the government is heavily involved in supporting research, either directly or indirectly, more still needs to be done for social work research to become fully effective. Each of the identified funding streams also has their own particular views on what they are willing to support and what they will not.

Social Work Publications

Once the research has been completed it then needs to be disseminated. Traditionally research has been published in peer-reviewed journals. In social work there is an ever-increasing number of these including:

British Journal of Social Work
Child Abuse Review
Child and Family Social Work
European Social Work

International Journal of Social Work
Journal of Social Work
Practice
Qualitative Social Work
Research, Policy and Planning
Social Work Education
Social Work and Social Sciences Review

There are also a growing number of social science journals that are not purely social work but cover interprofessional or related social work areas. These include journals like:

Age and Ageing
Journal of Interprofessional Care
British Journal of Criminology
Action Research
Disability and Society
Sociological Research Online
Health and Social Care

There is thus a wide and ever-growing range of outlets for social work research. It should though be remembered that the research reported in journals has often taken place 2–3 years previously. This takes into account the time to complete the research report, write the article, to submit it to a journal for peer review, to receive the article back probably with amendments to be made concerning any reviewers' comments, resubmit the article, the article to be accepted and then to await its turn for publication. Journals like the *British Journal of Social Work* are now providing an advance access to papers online before the journal is published due to the large number of accepted articles waiting their turn to be published. Most of the journals above are also published electronically; it is also likely that in the future all journals will be published online. There are also journals like *Sociological Research Online* and *Social Work and Society* that are solely published online.

Books are dealt with quite differently. Any book proposal is also peer reviewed but the difference here is not whether the work is good science or meets the required academic standards, but whether it will sell. Books, unlike journal articles, are peer reviewed in the proposal stage and then again after the book has been written. A social work book may be the most innovative research ever printed about that particular assessment technique or intervention, but if it is believed there is no market for it, in all likelihood it will not be printed. Like journals, books are increasingly being produced electronically and most libraries are now building up stocks of electronic books. How this will impact upon our buying of books and reading habits is yet to be seen.

Social work also has a number of regularly published 'trade magazines': *Community Care* and *Professional Social Work* being the most important in the UK. These trade journals contain job adverts, news items and short articles. Many of the articles besides containing commentaries on policy initiatives or current events will also report research information, which may be of use to the prospective researcher.

Social workers and students are increasingly expected to be able to access the World Wide Web (www) or the Internet. The Internet grew out of the ARPANET (Advanced Research Projects Agency Network) commissioned by the US Department of Defence for research into computer networking (Hewson et al., 2003). The amount of material available on the Internet is unfathomable and social worker, student or researcher needs to be careful to draw down materials of a high quality. Among the sites that you may want to use or select as favourites include government sites, for example, www.dfes.gov.uk or www.dh.gov.uk, social work research dissemination sites www.rip.org.uk and the Joseph Rowntree Fellowship section on research findings www.jrf.org.uk social work portals www.scie.org.uk, including its searchable database social care online or www.ssrg.org.uk, a helpful social care research and evaluation network with accessible online journal.

Besides the Internet, local authorities, trusts and many of the larger voluntary and independent agencies also have their own intranet system. Intranet systems usually only contain information accessible to that agency's employees and will cover the agency's major policies and procedures, key documentation and often other information of interest to employees. This is potentially a rich source of information for the social work researcher.

These various publishing outlets represent the major sources of written social work research although there is also an increasing use of videos/CDs of research material and poster presentations at conferences.

Social Work and the RAE – Assessing the Quality of Social Work Research

Guena et al. (1999) conducted an international comparison of the way 18 countries allocated research funding in higher education. The study identified four main means of research allocation: some form of research evaluation linked to funding as in the UK, Australia and Hong Kong; formula funding based on student numbers as in Germany, Finland and Denmark; negotiation between the government and the relevant research institution as in Austria or France; whilst

The Netherlands and the United States both have research assessment exercises but these are not linked to research funding.

In the UK the 2001 RAE (Research Assessment Exercise) was the fourth in the cycle of research assessment exercises that have provided an evaluative judgement on the quality of social work and helped to target HEFC research funding. The next RAE is to be held in 2008. This is an important event for social work and social work researchers when the credibility and standing of social work and social work research will be assessed and comparisons will be made between individuals, submitted research groups and between social work as a subject area and other subject areas.

The Discipline of Social Work

Reflexive Questions

Before reading any further I would like you to write down whether you believe social work is a discipline or not.

Why do you think that this is the case?

What subject disciplines like physics or history inform our understanding of social work?

It is contestable whether social work is a discipline or not. It can be argued that social work is not a discipline in its own right but an important area of activity studied from a variety of disciplines. Alternatively it could be argued that it is unclear what a university discipline is in the first place. At its most basic level a discipline could be defined as any individual subject area that a university wishes to define as a separate discipline. At the university level social work is located in a wide variety of departmental arrangements. For example, social work is located in the School for Policy Studies at the University of Bristol, in the School of Social Relations at the University of Keele, School of Social Sciences at University of Wales at Bangor and in the University of Salford, the largest single-sited social work qualifying programme in the UK, it is located in the School of Community, Health Sciences and Social Care. Social work did not appear in any of the four titles for the different locations of social work. This is not to say that there are not examples of social work being in the title of departments, for example, at Queen's University or the University of Glasgow.

Part of the explanation for this may be that it is difficult to describe anything as a distinctive social work discipline which does not include a range of other disciplines to add insight into the process. For example, at the start of this section you were asked to identify all the academic disciplines that you could identify that helped to inform social work. Your list probably has a number of the following academic subject areas:

- Social Policy
- Social Anthropology
- Sociology
- Psychology
- Social Psychology
- Law
- Philosophy
- Counselling
- Management
- Education
- Economics
- History
- Politics

It is also possible to add computing given that it is now a requirement of the social work degree that students are computer literate up to the European Computer Driving Licence (ECDL) standard or equivalent! Social work can thus be viewed as a value-driven activity in which various social sciences are brought to bear on the object of study that is social work. You may have also added mathematics given that social workers have had to increasingly be aware of the cost of residential placements or packages of care and the balancing of budgets.

At one level it is possible to argue that social work is a university discipline – there are groups of students studying in departments that have social work in the title. On the other hand it is difficult to conceptualise a distinctive subject area that is social work studies that is not informed by the other social sciences. This topic is discussed further in **Chapter 8**.

Summary

This chapter has covered a wide terrain generally subsumed under the notion of the business of research. The purpose of the chapter was to identify parts of the research process that are usually left unsaid in research text-books.

In particular we highlighted how research was funded, the disparity between health and social work research funding, research publication outlets and the disciplines that help to inform social work practice and research.

3 The Philosophy [of] Social Research

This chapter explores the major paradigms or ways of knowing about the social world. The major paradigms explored consist of positivism and interpretivism with an analysis of their associated research methods, strengths and weaknesses.

Positivism

Positivism has a long intellectual history dating back to at least Bacon (1551–1626) and Descartes (1595–1650). Auguste Comte is often considered as the first self-conscious voice proclaiming positivism. He argued that society, or the social world, could be studied using the same logic of enquiry as that employed in the natural sciences. Phenomena in both the natural and the social world were subject to invariant laws. The differences between them occurred because of their respective subject matters, which were little more than irritants to overcome by developing appropriate research techniques and methods. Importantly, such a view suggests a deterministic conception of the human race and society by effectively underplaying those factors regarded as uniquely human; free will, choice, morality, emotions and the like. In this approach the pursuit of knowledge is achieved through the process of induction which leads to experimentation, verification, explanation and finally to prediction.

Giddens (1977: 28–9) identified four major claims made by positivists:

- Reality consists of what is available to the senses.
- Science is the primary discipline.
- The natural and social sciences share a common unity of method.
- There is a fundamental distinction between fact and value.

The fourth of these factors suggests that facts, being the product of science, are superior whilst values represent an entirely different and inferior order of phenomena.

Positivism recognised only two forms of knowledge as having any claim to the status of knowledge: the empirical and the logical. The empirical is represented by the natural sciences and the logical by mathematics. The empirical is seen as the more important of the two.

The positivist view suggests that science represents the incremental unearthing of facts and laws, is progressive and in time all will be revealed. Empirical activity is based on the observation of 'brute data' that is, data that is not contaminated by interpretation or other subjective mental operation. In the same manner as natural scientists classify phenomena – by, for example, motion, shape, size and so on – social scientists should also classify the social world.

Associated with the positivist's view of classification is that of measurement. Accordingly, there have been great efforts to scale all kinds of variables in order to achieve an exactness and precision characteristic of natural sciences. It is possible to see this in many forms of psychology with its focus on personality correlates, stress or attitudinal scales.

Positivists believe the basis of science lies in the theoretically neutral observation language, which is both ontologically and epistemologically primary. As Delanty and Strydom (2003) note neither of these terms can be seen as unequivocal. It is though generally accepted that ontology relates to what is the essence of things that make up the world or the theory of the nature of the world. Epistemology, on the other hand, relates to what is the character of our knowledge of the world and what to count as facts. Both concepts are linked in that claims about what exists in the world almost inevitably lead to questions about how what exists in the world and how it is made known (Delanty and Strydom, 2003).

Statements made in the theoretically neutral observation language are directly verifiable as true or false by looking at the 'facts' of the world. This represents a correspondence theory of truth, in which the truth of a statement is confirmed by the correspondence with the facts. If it corresponds with the facts it is true, if not, it is false. Language and the facts 'would speak for themselves'. This form of empiricism can best be seen in the popular TV series C.S.I. in which Grisholm, a crime scene investigator, examines all the evidence, in minute detail, to ascertain the facts of the death and to discover whether it was murder and if so who the murderer was. Grisholm would famously let the 'facts speak for themselves'.

Similarly positivist statements would be directly verifiable as true or false by their correspondence with the facts. The beliefs we hold or the values we subscribe to are as factually 'brute' as atoms, velocities or simple harmonic motion. If social scientists or social workers would only use carefully constructed apparatus – questionnaires, likert scales and the like – inner mental states could, in principle, be researched empirically. Standardised lists could be developed – all social phenomena could be classified, correlated and measured. Hypotheses could be formulated and tested with a view to proving whether they were true or false. As such the world would become predictable and with predictability comes the potential for control. This would provide the social worker with the opportunity to intervene in social situations knowing that their interventions would result in positive outcomes – children would be safeguarded, people with disabilities would be empowered and the elderly supported to live their lives independently.

Falsification and Popper

Popper, a philosopher, is a key figure in the development of positivism who was particularly concerned not with how to verify a theory, which he believed was impossible, but how to refute one. Popper grudgingly admitted that metaphysical ideas may have helped with the development of science but that the primary purpose of empirical science is to draw a line between the empirical and the metaphysical science and pseudo-science. He developed the idea of deductivism or hypothetic-deductivism, logical reasoning. In it he contends that a scientific theory can never be accorded more than provisional acceptance (Popper, 1980). Popper gave the example of white swans to demonstrate his point. If our concept of a swan includes the notion of them being white we only need to see one black swan to falsify our previously held theory of what it is to be a swan. Popper was able to show that although deductivism was not able to prove a universal statement, in principle all statements remain refutable:

> There can be no ultimate statements in science: there can be no statements that cannot be tested, and therefore none that cannot in principle be refuted, by falsifying some of the conclusions that can be deduced from them. (Popper, 1980: 47, underlining in original)

Popper thus revised the orthodox positivist conception of science; no longer was the object of science to infer from specific instances to generalisations but to search for ways of refuting what he called 'conjectural hypotheses'. Science thus becomes not a body of accumulated and accumulating true theories but a series of conjectures or hypotheses that are yet to be refuted (Hughes, 1990). It is by critical trial and error that science progresses, only those theories that have passed the best tests available are hung onto, at least for the present whilst new tests are devised. The best theories will provide very precise predictions across a range of spheres allowing for empirical testing and opportunities for refutations. As such this suggests an almost evolutionary survival of the fittest.

Popper's contribution of positivism emphasises that theory acceptance must always be tentative. This can be challenged on both theoretical and practical grounds. On theoretical grounds if theory acceptance is always tentative, how can theory rejection be decisive when observation statements are by their very nature theory dependent and fallible? On the practical level Popper suggests social scientists should set out to disprove their theories, to find the 'black swan'. How realistic this is, is open to debate. From my own experience of researchers they often start with the opposite viewpoint and set out to prove what they already believe to be true. May (2001) also makes the point that if our empirical evidence falsifies a theory, is this sufficient reason for rejecting it? We may just have found a deviant case or new result that is yet to be built into the overall theory.

Reflexive Questions

At this point you should write down what you consider to be the key tenets of positivism and Popper's development of positivism.

Can you identify in what ways positivist assumptions are used in your practice?

The Interpretivist Reaction

Like the positivist tradition the interpretive tradition has its roots in the seventeenth century with Vico (1668–1744) who stressed that you could not study humans and society in the same way you studied inanimate nature. The former implied subjective understanding and thus required a wholly different method of inquiry to that of the natural sciences. Society, a product of the human mind was not only intellectually different, but also subjective and emotional requiring different models of explanation.

The alternative framework has continued to evolve stressing 'humanistic' or 'interpretive' approaches. These approaches have generally rejected the view that the scientific method could be applied to the study of social life and instead emphasised the importance of interpretation and understanding as the only legitimate ways of gaining understanding. The term interpretivism is being used here as an umbrella for a range of approaches including ethnomethodology (Garfinkel, 1967), phenomenology (Schutz, 1973), symbolic interactionism (Goffman, 1961), interpretive interactionism (Denzin, 1989), new paradigm inquiry (Reason and Rowan, 1981) and social constructionism (Berger and Luckman, 1979). In this section it is impossible to cover all of the nuances of the different approaches.

Nor would all of the proponents of each of the different approaches necessarily agree with this articulation of their position. However, as a starting point and for ease of understanding these approaches have been grouped under the interpretivist label.

The interpretivist tradition does not set out to gather facts or measure the frequency of occurrences. In fact, one of their major criticisms of the positivist is that they analysed out, or reduced to a set of statistics, those unique features which make social life a distinctive human experience. What exactly is left out is open to debate but may include choice, moral and political concerns, emotions, values or the self.

Schutz, an important interpretivist, expresses the difference between the natural and social sciences like this:

> The world of nature, as explored by the natural scientist, does not 'mean' anything to the molecules, atoms and electrons therein. The observational field of the social scientist, however, namely a social reality, has a specific meaning and relevance structure for the human beings living, acting and thinking therein. (Schutz, 1978: 31)

Thus one of the key differences between positivism and interpretivism can be seen to be the ability of the subjects of social life to create their own commonsense structures and to be able to interpret their own experiences. In other words, unlike atoms or molecules, social actors can talk about, explain to others or justify their actions. Knowledge is then not something 'out there' to be discovered, but something derived and created from the experiences of the social actors.

This highlighting of meaning and the describing of actions leads the researcher to impute motives of one sort or another. The analytic force of motives and reasons lie not so much in their being 'internal' but in their being tantamount to rules for governing behaviour. This behaviour though can often be explained in a variety of ways. Thus it is possible for a social worker with a 'nod' to communicate to a service user they wish them to continue with what they are saying, to indicate they agree with what is being said, to indicate they disagree with what is being said or to indicate to a colleague they want to leave. The same movement can in various circumstances and with various intentions constitute any of these actions; it could also be argued that as a result of this it constitutes none of them. The observer cannot see directly into the minds of the observed to inspect their motives. Nonetheless if certain particulars of the context are supplied an interpretation, of greater or lesser value, can be attempted.

Human action arises from the sense that people make of different situations. The positivist tradition constructed their version of social reality by drawing a distinction between identifiable acts, structures and institutions as 'brute' facts, on the one hand, and beliefs, values and attitudes on the other. These two orders of reality were then correlated in order to derive generalisations or regularities which then become the substance of social life. The 'brute' facts are considered as objective whilst the values, beliefs and attitudes are considered as a subjective reality, an inferior status. The elements of meaning were thus relegated to secondary versions of reality.

For the interpretivist reality cannot be identified apart from the language in which it is embedded. Social realities are constructed, re-constructed, negotiated and re-negotiated in and through meanings. Meaning is thus not only about grammatical rules, but is also about social interaction. Language and the importance of language is important in this tradition. The reality of the natural or social sciences cannot be known independently of the concepts available in language. Secondly, meanings are not totally idiosyncratic – otherwise it would be impossible to communicate. This is not to deny that there are differences in the way that black and white people, women and men, the disabled and non-disabled or children and adults all experience the world. Meanings are therefore not finitely specific, but achieve meaning from their background, context and the interpretations of the language speakers and receivers. Finally, disputes about meaning do not necessarily stem from deficiencies or inadequacies of natural language but may represent inherent features of social reality.

Interpretivism and relativism

The interpretivist cannot help but be continuously engaged with his/her own subject matter. This poses the eternal problem for the social researcher in how they can maintain that their reality or interpretation is more accurate and valid than that of the subjects of the study. This is particularly the case when we remember that both use the layperson's world as their reference point and share the same resources.

Hammersley (1995) identifies relativism as the key stumbling block for interpretivists. Whilst accepting that there is no universally agreed definition of relativism, what appears to be central to the notion is a view of knowledge as paradigm dependent. Paradigms are discussed later, but it is useful to think of them here, as a worldview containing a set of assumptions that go beyond rational explanation. When someone makes a knowledge claim we can ask by what criteria can he or she judge his or her knowledge claim to be true. Then, when the criteria are presented we can ask on what basis this set of criteria is believed to be valid. When that basis is identified we can then ask why that is believed to be true. The argument becomes circular with no definitive end point. The interpretivist becomes hoisted by their own petard, not only is a relativist position false, when viewed from other positions, but it also is within the framework of interpretivist assumptions.

In the end we need to recognise that absolute certainty is not available, and that attempts to produce absolute knowledge by sense data are bound for failure. Such a view is in danger of sinking into total relativism wherein anything can be true in some framework, if not our own. To avoid this position Hammersley (1995) invokes Pierce's 'commonsensism', where assumptions are relied on until subject to genuine doubt. This potentially creates more difficulties than it answers, as there is often nothing common about commonsense. One group's taken for granted assumptions do not necessarily transfer to any other group. In the 1990s Oliver (1990, 1993) argued that people with disabilities should not take part in research unless it was informed by the social model of disability as their experience of able-bodied researchers commonsense understandings were experienced as oppressive and discriminatory.

At this point it is worth mentioning that both positivism and interpretivist paradigms can be criticised for being inherently conservative and failing to deal with issues of power. A full discussion of this critique is not possible here but students should be aware that there are more critical research approaches and more committed research strategies some of which are discussed in **Chapter 8**. We next move onto research methods or the tools that researchers use to help them make sense of their data.

Reflexive Questions

Having read this section on interpretivism you should write down what you consider to be the key tenets of interpretivism and its critique.

Can you identify in what ways interpretivist assumptions are used in your practice?

Having looked at both the interpretivist and positivist approaches which do you favour, and why?

Research Tools

Having read the preceding sections you will have noted that both positivism and interpretivism make different assumptions about how the world can be known. Before we look at some of the well-known research methods it is important to note that new research methods are continually under development and there are a wide range of both qualitative and quantitative methods that social work researchers could use including documentary analysis, discourse analysis, conversation analysis, biographical methods and increasingly complex statistical quantitative methods. There is also a growing interest in the use of computers, drama, drawings, text and other mediums for research exploration. These methods and mediums are beyond this text-book but the interested reader needs to be aware that there is a wide variety of approaches available for research exploration. You are now invited to consider which of these more common methods you would associate with which philosophical approach.

Reflexive Questions

Can you identify which of the following you would associate with a positivist position?

1 Participant observation
2 Questionnaires
3 Surveys
4 Interviews

What reasons did you give for your view?

Traditionally research methods are split between quantitative and qualitative approaches. Positivist researchers are often associated with quantitative methods including random control trials, surveys and questionnaires. It is important first to clarify what we mean by quantitative:

> Quantitative methods (normally using deductive logic) seek regularities in human lives, by separating the social world into empirical components called variables which can be represented numerically as frequencies or rates, whose associations with each other can be explored by statistical techniques, and accessed through researcher-introduced stimuli and systematic measurement. (Payne and Payne, 2004: 180)

'Quantitative methods' is an umbrella term and covers a wide range of methods that have been informed by positivist assumptions. At its simplest it involves counting how frequently things happen (for example, the number of GCSEs a child looked after attains) and the presentation of these frequencies in tables and graphs. This is then often extended to look at whether two or more factors are associated, related or can even be seen to be causal. In the example above Jackson (1987, 1994) and Heath et al.

(1994) have all demonstrated that being accommodated is associated with poor educational attainment which is also associated with poor life chances. This is one of the reasons why the Government in its performance assessment framework, set educational attainment targets for local authorities to achieve for children looked after.

Payne and Payne (2004) identify the common features shared by almost all forms of quantitative research.

> - The core concern is to describe and *account for regularities in social behaviour.*
> - Patterns of behaviour can be *separated into variables, and represented by numbers.*
> - Explanations are expressed as *associations* (*usually statistical*) *between variables,* ideally in a form that enables prediction of outcomes from known regularities.
> - Social phenomena is explored through *systematic, repeated and controlled measurements.*
> - They are based on the assumption that *social processes exist outside of an individual actor's comprehension,* constraining individual actions, and accessible to researchers by virtue of their prior theoretical and empirical knowledge. (Payne and Payne, 2004: 181–2, italics in original)

In seeking to explain quantitative research methods I would like to examine random controlled trials, surveys and questionnaires.

Random Controlled Trials – 'The Gold Standard'

The random controlled trial (RCT) is most closely associated with medical research, but it is possible to undertake RCTs in other knowledge domains. RCTs are also closely associated with a positivist paradigm of research knowledge.

The RCT is often cited as the 'gold standard' method of assessing the efficacy of treatment methods (Reynolds, 2000). The central feature of an RCT is the random allocation of potential participants to an experimental or control group. The intention is to eliminate bias by ensuring all conditions are randomised including the preferences and expectations of patients and doctors. In some of these trials the treatment group will be further divided with one half of the group receiving the treatment and the other half a placebo. Ideally this is done as a 'double blind' where neither the patient nor the doctor are aware of who is receiving the treatment and who is not. Similarly, whoever is evaluating the clinical outcomes should not be aware of who has and who has not been treated.

RCTs then follow up their patients to identify the relative outcomes for the control and experimental group with a view to establishing whether the results are clinically important or not. Clinical importance is often used to replace statistical significance. Clinical importance may refer to the speed of recovery of the patient, whether the patient has suffered side-effects, required re-admission to hospital or even whether the patient survived the treatment or not. Clinical importance though may be identified differently by the anaesthetist (the patient survived the operation), the surgeon (the patient was discharged from hospital), the GP (the patient was still alive/survived six months after the operation) and the patient whose quality of life had improved.

Sheldon (1986) and MacDonald (1996) have reviewed the relatively few RCTs in social work and have argued that more are needed. The question of RCTs and social work is further discussed in **Chapter 5** on evidence-based practice.

Surveys – 'Would you mind if I asked you a couple of questions?'

All of us will have experienced being stopped in the street by someone holding a clipboard who then asked us a number of questions about which adverts we watched, drinks we drank or foods we ate. During this process the researcher would be ticking boxes in response to our answers. While this is market research, the use of surveys is also a central part of social research. Surveys are one of the most frequently employed methods in social research. Governments, academic researchers and campaigning organisations alike, use surveys (May and Williams, 2001). An example of this is the British General Household Survey which covers a wide range of socio-economic data including: family type, car ownership, occupation, religion, education etc. and occurs every 10 years and is expected to be completed for every person over 18. This allows the researchers not only to track the changing nature of individual households, but of British social life in general.

Ackroyd and Hughes (1981) have characterised surveys under four different headings.

- The factual or social survey aimed at eliciting general facts, rather than opinions or attitudes, about the conditions and the organisation of whole societies.
- Attitude surveying which focuses on the attitudes people hold as a means to seek to explain and potentially predict their behaviour. This represents a move away from factual surveying to public opinions.
- Social psychological surveying which is more explanation and theory oriented and seeks to investigate personality via various types of attitude measurement techniques.
- Explanatory surveys designed to test some theoretical explanation.

The first two of these were designed to achieve practical as opposed to theoretical ends, although it could be claimed that they all seek to provide or contribute to some degree of explanation.

Payne and Payne (2004: 219) note that surveys have typically three types of characteristics:

> They *collect data in a standardised way* from *a sample* of respondents, enabling the data to be *codified, normally into a quantitative form.*

There are an increasing range of surveys and these include face to face interviewing, telephone, postal, e-mail or texted surveys. As can be seen some of these survey types require the interviewer to be present and others are self-completion. Where the interviewer is not present to explain the questions it is very important for the questions to be clear and unambiguous. Often a survey will be piloted to pre-test the questions for their ease and clarity of operation, to check whether they are addressing the research question(s) and whether they provide a means of differentiating between different respondent groups.

The survey questions are usually contained in a questionnaire, which starts with a set of classificatory questions contained in the 'personal' section of the questionnaire – these may include questions about age, gender, ethnicity, occupation, salary etc. The researcher must also make a judgement whether to use open or closed questions. Open questions allow the interviewee greater freedom to respond to the question in a way that suits their interpretation. Closed questions limit the number of possible answers allowing for easier and cheaper analysis.

Surveys very rarely report a 100 per cent response rate, for example, people move home or go on holiday and cannot then be traced, or die. Generally 70 per cent is seen as an adequate response rate for face-to-face interviews. In self-completion and postal surveys 33 per cent is seen as more typical.

Easterby-Smith and Thorpe describe the benefits of quantitative methods in general and surveys in particular as:

> They can provide wide coverage of the range of situations; they can be fast and economical; and, particularly when statistics are aggregated from large samples, they may be of considerable relevance to policy decisions. On the debit side, these methods tend to be rather inflexible and artificial; they are not very effective in understanding the processes or the significances that people attach to actions, they are not very helpful in generalising theories. (Easterby-Smith and Thorpe, 1996: 32)

Qualitative Methods

The preceding two research tool methods were primarily associated with quantitative methods – the next three with qualitative. Qualitative methods are linked to the interactionist perspective of philosophy. Silverman (1993: 170) has stated that qualitative methods are 'especially interested in how ordinary people observe and describe their lives'.

> Qualitative methods produce detailed and non-quantitative accounts of small groups, seeking to interpret the meanings people make of their lives in natural settings, on the assumption that social interactions form an integrated set of relationships best understood by inductive processes. (Payne and Payne, 2004: 175)

Qualitative, like quantitative is an umbrella term. Qualitative methods refer to a set of approaches that share common features. Qualitative methods:

- Focus on seeking out and interpreting the meanings that people ascribe to their own actions.
- Actions are seen as contextualised, holistic and part of a social process.
- Seek to encounter social phenomena as they naturally occur.
- They work with smaller samples looking for depth and detail of meaning with a less general and abstracted level of explanation.
- They use inductive as opposed to deductive logic allowing ideas to emerge as they explore the data. (Payne and Payne, 2004: 175–6)

Thus the qualitative methods focus on individuals, their interactions, emphasising interpretation and meaning and the ways in which mutual understandings are negotiated. This alternative research direction requires a different set of methods than that of the quantitative researcher. RCTs or surveys with their emphasis on statistics are unable to capture the detail of social life requiring different techniques and different methods.

Qualitative methods have the ability to look at change processes over time, to understand people's meanings, to adjust to new issues and ideas as they emerge. They also contribute to the evolution of new theories and are seen as 'natural' rather than as 'artificial'. In particular we are going to focus on participant observation, ethnography and interviewing.

Participant Observation

Participant observation requires the researcher to immerse themselves in the lives of those being studied. Participant observation thus requires the researcher to engage in a number of activities including looking, listening, enquiring and recording. May (2001) comments that the apparent naturalness of this approach leads those new to social research to assume they can undertake this approach with ease. On the contrary it is probably one of the most personally demanding and analytically difficult methods of social research to undertake. Depending on the aims of the study the researcher may be required to spend a great deal of time in unfamiliar surroundings, building and maintaining relationships with people with whom they have little personal affinity; making copious notes on what to others would appear mundane happenings; putting themselves at a degree of personal risk and then spending months analysing the data after the fieldwork has been completed.

There are two main types of participant observer, the researcher as 'complete participant' and the researcher as 'participant-as-observer'. In the 'complete participant' role the researcher becomes a fully-fledged member of the group under study, the purpose of the research being concealed from the group under study. There have been a number of such studies in the past including one where a social work lecturer caused a major stir by living in a council estate in Glasgow whilst she observed and wrote down her views of lifestyle of the residents on that council estate. Following publication of her views some of the residents were very disparaging about her comments and felt let down that she had written such comments about them. In this type of approach the residents were not aware they were being observed nor of the value judgements that were being made about them and the way they lived.

The more common approach is of 'participant-as-observer' wherein both the researched and the researcher are aware that theirs is a fieldwork relationship. Since the research relationship is not concealed the researcher is able to use other methods, for example interviewing to complement their research. This type of research has been used with teenage gangs, drug users and in residential settings.

Participant observers may work in teams, but more often they work alone. In the process of observing they witness how their research subjects interact with their social environment continually interpreting and applying new knowledge. The researcher or ethnographer is the instrument of data collection. The ethnographer enters the research subject's social universe using a range of techniques that could include 'living among people', interviews or life histories. Researchers accept that they will 'contaminate' the situation although in doing ethnography this engagement is used to an advantage. In this process ethnographers have drawn upon their own experience and biographies to help understand the research process. This use of the researcher's own cultural equipment is used reflexively to make sense of social action in context. Reflexivity thus implies that 'knowledge is made rather than revealed' (Taylor and White, 2000: 199). Ethnography thus requires us to consider how power is exercised in the research process and the implications of this for what does and does not constitute knowledge.

Focus Groups

In recent years there has been an explosion in the use of focus groups for all sorts of activities including the market testing of new products, voting intentions and within a wide range of social care activities where service user perceptions are wanted to help inform service provision. Cronin (2001: 165) defines a focus group as a group interview:

> A focus group consists of a small group of individuals, usually numbering between six and ten people, who met together to express views about a particular topic defined by the researchers. A **facilitator,** or **moderator,** leads the group and guides the discussion between the participants. In general focus groups last one and a half to two hours and are tape-recorded. (bold in original)

Focus groups are thus managed discussions that are organised to explore a specific set of issues that involve some kind of group activity. Focus groups are particularly useful when the researcher wants to explore people's experiences, opinions and concerns. What sets them apart is that focus group participants, unlike individual interviews or questionnaires, engage in discussion with each other creating an interactive and dynamic process led by a moderator or facilitator.

There are at least four advantages of using focus groups:

- They provide an opportunity to observe and collect a large amount of data and interaction over a short period of time.
- Discussions should provide rich data as participants present and defend their own views whilst challenging the views of others.
- This very process may help participants clarify their own views but also open them up to alternative views they would not have considered.
- Focus groups encourage theorisation and elaboration.

On the negative side is a concern that the data focus groups generate is limited in its generalisability. This is because focus groups do not generally consist of randomly selected individuals and those who are willing to participate in a focus group may be different from those who are not. There is also an issue of interviewer bias whereby the researcher is required to be both reflective and reflexive in order to minimise the impact of their role.

Due to these difficulties focus groups are often included as part of a research strategy although they can be used and justified in their own terms.

Research Techniques can be both Quantitative and Qualitative

In this section I would like to look at a major research tool that can be both quantitative and qualitative. The importance of this method is that it is closely related to the key social work skill of interviewing. Traditionally the worlds of the social worker and the researcher have been seen as mutually exclusive and esoteric activities. One sat in an ivory tower theorising about the world whilst the other had to deal with the messy problems and business of the world. Both, though, may have more in common than is usually acknowledged. I would like to suggest that interviewing is a central technique for both. Kadushin (1972: 8) in a classic text on interviewing described the essential features of an interview as:

> a specialised form of communication. A communication interchange in the interview involves two people, each of whom possesses a receiving system, a processing system and a transmitting system. The receiving system consists of the five senses, the receptors. Communication in the interview involves primarily the use of two sense receptors – the eyes and the ears. Having received the incoming signal, one processes it; this involves making sense of the received message, giving it meaning. The processing activity consists of recalling stored information, relating related information to the message, thinking about the message, evaluating the message, translating it so the message is coherent with the receiver's frame of reference. As receivers we select certain items from the incoming message, ignore others, and rearrange what we hear into interpretable patterns. We then typically formulate a response.

Interviewing aims at establishing a framework for future evaluation and enquiry in both spheres. Both probe and must be continually aware of the conditions under which information is generated. The difference occurs at a later stage when social workers use the data from the interview to assess a situation and to decide whether an intervention is required, and if so, what type of intervention. This does not mean social workers eschew theoretical considerations or give up their analytical powers but that their priorities are different. It should also be remembered that researchers do not necessarily stop at the theoretical, but may wish to see their work established as current practice or result in a change of policy.

At this point it could be suggested that interviews could be seen as a qualitative tool given their negotiative and social constructivist agenda. As the reader will already be aware such distinctions are not always as clear-cut as we would first believe. It is possible for an interview to be quantitative, qualitative or both. What decides whether it is one or the other is the nature of the questions being asked. When we discussed surveys we noted that the questions could be open or closed. Quantitative interviews would use closed questions and qualitative interviews open questions. One would be asking the respondent to identify a predesigned response whilst the other would use a much more discursive approach, possibly probing answers to elicit deeper and more meaningful understanding.

An interview is neither inherently qualitative nor quantitative although we normally associate it with qualitative approaches. Interviews also regularly use schedules to help direct the researcher to the question areas in which they want a response. Within the social sciences semi-structured interviews are very common which include both closed (quantitative) and open (qualitative) questions.

The closed questions may be seeking to gain comparative views of a particular policy of implementation strategy and the open questioning trying to help understand the reasoning behind a particular view. The closed questions may focus on outcomes, but without the open questions they suffer the severe limitation that they do not tell us what these outcomes were outcomes of.

This is an important point for us as social workers and researchers – we need to continually question the assumed, we need to ask ourselves who says something is true and, importantly, what evidence have they got for believing it is so. These are issues we will visit in more detail in **Chapter 5** when we discuss evidence-based practice.

In the last two sections we have looked at some examples of qualitative and quantitative research methods. At one level it has been possible to associate different methods with different research epistemologies. However, a commitment to qualitative approaches does not necessarily imply innumeracy as qualitative methods often include statements about sample proportions and analysis of field notes may include content counting. It has also been possible to show that research methods can be both quantitative and qualitative, or even both at the same time. At this point it would be useful to explore the notion of a paradigm and how this impacts on how the world can be known.

Paradigms – as Worldviews

It is often presented that the positivist and interpretive perspectives on the world represent different paradigms of how we understand the world. Paradigms are generally viewed as important constructs following the publication of the *Structure of Scientific Revolutions* by Thomas Kuhn in 1962. Kuhn, a historian discussed how natural scientists engage in debates about the phenomena they study and how they move from one major theory to another.

Kuhn (1970: 175) defined paradigms as 'the entire constellation of beliefs, values, techniques and so on shared by members of a given community'. As such paradigms are the worldviews or theories, which define and legitimate problems, methods and solutions for a scientific community. While Kuhn's views were developed in response to the natural sciences his views are also seen as very pertinent to the social sciences. As Gilbert (2001) acknowledges, social research (and we can also include social work research in this) is situated within a 'paradigm', a scientific tradition. Any new research project is linked into what has gone before. The problems any social work researcher will tackle are derived from previous social work research, will have been discussed in the relevant journals and the methods used will have been honed by previous researchers. Evidence of the linking between new research to previous ideas and concepts is an important function of the 'references' sprinkled throughout the article or book. These references not only acknowledge previous work, but also borrow their authority and that of the author to legitimate their own work.

Gilbert (2001) suggests that knowledge is constructed and linked to a particular professional knowledge community. This links back to Kuhn (1970) who suggested that during periods of 'normal science' the focus is on first, clarifying the facts within the paradigm, secondly, comparing the facts with predictions from the paradigm and thirdly, articulating and fleshing out the paradigm. During this period of normal science if results do not fit into the theory they are overlooked, suppressed or explained away as a novelty. With time these novelties come to be seen as more than just another puzzle requiring solution. These novelties become too great to be ignored and there begins a transition to crisis and abnormal science when the very foundations of the science become challenged.

This creates the conditions for a change in paradigm choice. Kuhn (1970) claims the choice of a paradigm is not a rational process and can never be settled unequivocally by just logic or experimental evidence. The new paradigm is not a special case of the old one, but is incompatible and incommensurate with what has gone on before.

Paradigms and Incommensurability

A critical concept in relation to paradigms is the notion of incommensurability. The traditional view of incommensurability is that rival positions, or paradigms, are irreconcilable. It is not possible to believe in both positions at the same time, to believe in one is to deny the other. They represent competing worldviews, with differing assumptions about the nature of reality and how it may be known. The second aspect of incommensurability relates to the relationship between paradigm and method. As previously noted in this chapter, both positivism and interpretivism helped to inform particular social research methods. Research incorporating quantitative methods presupposes certain beliefs about what the world is and how it can be known. Likewise, research with qualitative methods makes different competing and antithetical assumptions. Paley questions whether this relationship is either logical or pre-ordained in asking: 'why should anyone feel the need to say that methods, as such, presuppose anything?' (Paley, 2000: 148).

In order to make his point Paley usefully draws on the metaphor of maps, in particular a motorway map and an Ordnance Survey map. It is tempting to suggest that an Ordnance Survey map is more detailed and complete than a motorway map. But, for the motorist the fact that a motorway map does not include every bridle path or footpath does not mean that it is incomplete. Indeed, for the motorist the fact that the map is not to scale is a benefit; otherwise it would be impossible to see the roads. This he suggests is a matter of function, and methods like maps are dictated not by inherent philosophical positions but by particular functions. Both maps are tools and are only suitable for particular purposes and tasks. What then should drive a choice of method is not a philosophical standpoint position but a question as to what is the most appropriate tool for the task in hand. Certainly the viewpoint of this book is that it is more important to ensure a suitable tool between the research questions than the degree of philosophical coherence of epistemological positions traditionally associated with particular research methods. This also opens up the possibility of being able to use different methods in the same study – the motorway map of survey approach to gain a broad bush understanding of a particular issue, for example the characteristics of children in residential care including their age on admission, reason for admission, length of time in care, number of placements etc. This could then be followed up with an ethnographic study to examine the meaning for being looked after as experienced by young people, their carers and their parents. As Ritchie (2003:43) observes:

> When using qualitative and quantitative research in harness, it is important to recognise that each offers a different way of knowing about the world. Although they may well be addressing the same research issue, they will provide a different 'reading' or form of calibration on that issue. As a consequence, it should not be expected that the evidence generated from the two approaches will replicate each other. Instead the purpose of interlocking qualitative and quantitative data is to achieve an extended understanding that neither method alone can offer.

Thus when using quantitative and qualitative research methods it is important to recognise that not only has each a different way of knowing about the world, they will each also approach the question differently and each provide a different answer. As a consequence it is imperative for researchers who use mixed methods approaches to explain why the data and their 'meanings' are different and to avoid the situation where one approach becomes dominant and conflicts between the data become hidden.

Triangulation is potentially important here – triangulation originated from the quantitative research 'multiple operationalism' of Campbell and Fiske (1959) whereby multiple measures are used to ensure that the variance reflected is that of the trait or treatment and not associated with any other measure. Triangulation has also come to mean 'convergence' between researchers and convergence amongst theories. To achieve this it is generally accepted that the researcher should pick triangulation sources that have different biases and different strengths so that they can complement one another. Triangulation thus involves the use of different methods and sources to check the data's integrity and or to extend the inferences that can be drawn from the data.

We should, though, not just assume that multiple methods, or the use of triangulation, would automatically lead to sounder conclusions. Shaw (2003: 110) approvingly quotes Trend's classic account of an evaluation of a USA programme evaluating the effectiveness of direct payment of housing allowances to low income families. Trend concluded:

> The complementarity is not always apparent. Simply using different perspectives, with the expectation that they will validate each other, does not tell us what to do if the pieces do not fit. (Trend, 1979: 110)

Using multiple methods and triangulation does not remove the responsibility from the researcher to ensure that these methods work together in such a way that they add additionality and address the research question. Using triangulation does not remove the responsibility for ensuring that generated data is analysed rigorously and methodically identifying both areas of correspondence and dispute. Both aspects need to be attended to.

MacDonald (1999: 98) argues that if there is to be a rapprochement of quantitative and qualitative approaches four principles will need to be conserved:

- as far as possible, researchers should make explicit the assumptions and values underpinning the questions they ask and the methods they deploy.
- the methods should both be internally and externally robust and valid.
- the methods employed should neither be oppressive to the researcher or the researched.
- that the research should be oriented towards knowledge that can be used by users themselves.

Whilst the first two principles would appear to be supportive of Paley's position the second two are of a different order. The last two principles could be seen as particularly important to ethically driven professions such as social work with a value base and commitment to service user empowerment.

Paley's perspective is not universally accepted and D'Cruz and Jones (2004) argue that you cannot ignore that methods have a relationship to the philosophical positions in which they have their roots. They do though argue for a more pragmatic approach to method choice to ensure powerful groups will engage with the results and not just seek to attack the methodology. In so doing they note that discussions about the relationship between paradigms and methods has become more relevant as knowledge and social research has challenged the previously unseen positivistic, scientific, white, Western, male perspectives. Feminist, postmodern, post-structural and post-colonial perspectives have all served to reinforce the positioning of the knower as essential in the creation of knowledge and given marginal groups a voice. This political standpoint position suggests that the choice of methods is not just an issue of the appropriate tool for the task, but also a statement concerning how the researcher positions him/herself and how he/she considers the world can be known. This does not mean that we should adopt an incommensurate or different paradigm position, but demands that we make clear the intellectual, ethical and methodological assumptions with which we are working. D'Cruz and Jones (2004: 57) express their position well:

> as social workers, we must be aware of the political and ethical processes of knowledge construction. Social work research is just one of these aspects. If we are to achieve social change as social workers, we are immediately positioned both politically and ethically in relation to social issues and social problems. Therefore we cannot escape our personal or professional assumptions or goals.

At this point it is important to understand where the writer locates himself. I am a white, middle-class (at least by occupation, if not by birth) male, married with two children, brought up in Northern Ireland during the 'Troubles', but who has not lived there since finishing his first degree. Politically, I adopt a critical perspective and see the world as an unequal place both within the UK and between the 'developed' and 'underdeveloped' worlds. As a social worker, or social work manager, for most of my career I have struggled with the contradictions and ambiguities of social work practice as I strived to steer a course in which service users could be empowered, or at worst, be no worse off after than before my contact. As I have predominately worked in childcare this has created difficulties with the needs of children and parents not always being conterminous. On occasions, this has also included differences in interests between brothers and sisters and mothers and fathers. I believe social work can make a difference, but is in danger of becoming overly administrative and losing the human connection and 'subversive' nature which gives it its critical edge. I hope this will help the reader to begin to identify some of my biases and be able to read this book with the knowledge of where the writer is located and ask themselves how would a writer located in a different standpoint respond to the issues identified in this book.

Reflexive Questions

How would you locate yourself?

How does this impact upon how you see the world?

How does this relate to why you want to be a social worker?

How does this impact upon the area of social work you want to follow?

Summary

This chapter has raised a number of important philosophical issues that you may feel the need to re-read to begin to understand fully. Often social workers and researchers find philosophy difficult preferring to be 'doers' rather than 'thinkers'. But, as I hope, you now appreciate both concepts are linked and how you think impacts on what you do and vice versa. In particular the chapter has highlighted two of the major paradigms of social research – positivism and interpretivism, identifying some of their key assumptions and how both approaches represent different versions of how the world can be understood. The chapter has also identified some of the key social research methods and shown how they relate to the different paradigms. This was followed by a discussion on the nature of paradigms, their incommensurability, and whether this incommensurability also relates to method. In response to this we noted how some researchers will only use one type of method, but have suggested a way in which the nature of the research question may drive the research method used. Even so, it has been suggested that the researcher also needs to reflect on how their own standpoint will affect what methods are chosen and how they are combined. The researcher's standpoint was seen to be very important and we will return to this later when we discuss service user research in **Chapter 6** and anti-oppressive research in **Chapter 7**.

Suggested Reading

D'Cruz, H. and Jones, M. (2004) *Social Work Research: Ethical and Political Contexts*. London: Sage. This is a well written, easy to read book which was also identified in **Chapter 1**.

Hughes, J. (1990) *The Philosophy of Social Research* (2nd edn). London: Longman. A challenging but engaging introduction to the philosophy of social research.

May, T. (2001) *Social Research: Issues, Methods and Process* (3rd edn). Buckingham: Open University Press. This book provides an accessible account of key issues in research.

Payne, G. and Payne, J. (2004) *Key Concepts in Social Research*. London: Sage. This book provides short accounts of key concepts whilst also directing the reader to further studies and descriptions.

4 Ethical Issues in Social Work Research

This chapter examines the ethical issues surrounding the undertaking of social work research. It begins by identifying the rationale for ethical research behaviour and then discusses the nature of ethics and social work as a basis for a social work and social care ethical research code. This is then followed up by a discussion of ethical issues as related to the research process and is divided into issues before the research commences, during the research and after the research has been completed. This chapter takes the perspective that ethics is not a 'bolt on' but an inherent, demanding and ongoing part of the research process (Humphries and Martin, 2000).

Ethics, in its grandest form, may be recognised as a philosophical discipline whose primary concern is the science of morality. Ethics though takes on a more distinctive form when applied to social research. Here ethics refers to the standards established within the social work research profession for the conduct of its members (Homan, 1991). It is with this conduct that this chapter is concerned.

The Rationale for Ethical Research

It could be argued that there should be no concern about the nature of ethics of social work researchers as many are qualified social workers for whom ethical issues are integrally entwined at the heart of social work practice. We have already discussed that the nature of research is neither intrinsically neutral nor necessarily beneficial. This is most graphically seen in medical research. The Tuskegee Syphilis study between 1932–72 resulted in the denial of treatment of those with syphilis to allow researchers to study and chart the progress of the disease to its end point, the death of the respondents (Brandt, 1978). Live cancer cells have also been injected beneath the skin of non-consenting adults to chart the progress of the disease (Barber, 1976). In New Zealand, women with pre-cancer symptoms were assigned to one of two groups without their consent. One group received treatment, the other did not (Smith, 1999). These examples not only highlight dominant discourses and power dimensions as to which groups can be tested for the 'greater good', they also challenge their commitment not to tolerate discrimination of any form and to seek to promote emancipatory research as suggested by the ethical code for social work and social care. Recently, in the UK there has also been the organ retention scandal at Alder Hey Hospital where dead babies were kept for bio-medical research without the prior consent of their parents, all of whom thought that they had already buried their children, organs and all.

These examples from bio-medical research are important – as we shall see, the attempts to control bio-medical research and protect its subjects has been a major driver for changes in the social sciences. There are also many examples from the social sciences that indicate the need to protect research subjects. One of these is detailed below and others will be introduced as the chapter progresses.

Homan (1991) views Humphrey's study of the 'tearoom trade' as having the most unethical methodologies in the history of social research. The 'tearooms' in which Tearoom Trade (Humphreys, 1975) took place were men's public toilets in the United States. Humphreys assumed the role of 'watchqueen' or lookout who was expected to cough when strangers appeared thus alerting those men who were engaging in physical interaction which he observed and later noted down. Humphreys' records included the details of the men's age, dress and car registrations through which he was later able to trace the men to their home addresses. Humphreys was also involved in a health survey and included some 50 of his own subjects whom he later visited at their homes after waiting a year and changing his dress and hairstyle. He was satisfied that the men did not recognise their former 'watchqueen'. Homan (1991) quotes Warwick who commented that 'The concentration of misrepresentation and disguises in this effort must surely hold the world record for field research' (Warwick, 1982: 46).

In response to these concerns about ethical research practices differing professions have developed their own 'code of ethics'. These include the Social Research Association (www.the-sra.org.uk), British Psychological Association (www.bps.org.uk) and British Sociological Association (www.britsoc.co.uk). Butler (2002) has made the argument for a distinctive code of social work and social care research. He recognises that any such code is a form of professional claims-making and represents a measure of self-interest and as such the relationship between those who wish to be included in the code can be as important as the code itself. It is thus important that ethical codes are contextualised and situated. This leads us to two important implications. First, codes of ethics, like the moral principles on which they are based, are normative. Secondly, such codes can never be morally and ethically neutral. They inevitably reflect and articulate the occupation's ideological and moral aspirations (Butler, 2002).

If social work research is to claim its own distinctive ethical research code it can only do so by claiming disciplinary or occupational distinctiveness. As such this argument proposes that social work and social work research occupy the same discursive site and generally serve the same audiences and fields of interest:

> Social work research is about social workers, what they think, what they believe, what knowledge they claim and what they do with it and its primary (but not its only) audience will be social workers, service users and those who determine who falls into which category for the purposes of public policy. If this is so, then the ethics of social work research must logically be at least compatible if not coterminous with the ethics of social work generally. (Butler, 2002: 241)

As Butler (2002) himself admits such a proposition is not without its difficulties – first, terms like 'social work' and 'social worker' are contested and contestable and are located within particular temporal and spatial contexts. Secondly, this characterisation of social work research may be interpreted as implying that only social workers can undertake social work research. Given the recruitment criteria for employing social work academics and researchers this is likely to be true but is not necessarily so. Of more interest is the case of other professional groups, for example, psychologists or management researchers undertaking social work research into, for example, the levels of stress of the field social workers or into the effectiveness of residential care management. In these cases should their research be rooted in the ethics of social work research? Butler (2002) argues that the answer is 'probably not' if the research was related to their own fields of interest, discourses and usual audiences. However, he does not identify when it should be rooted in the ethical code of social work research. Of the two examples identified it could be argued that if levels of staff stress were being researched using a questionnaire and neither researchers nor workers ever met together the acceptance of a social work code of ethics would not be necessary. The psychologists own code of ethics would be suitable to safeguard staff in this situation.

The second example is potentially less straightforward. Let us say the management researchers were examining management effectiveness in a therapeutic community of young children. As such they wanted to undertake participant observation of managers and workers in situ and also wanted to interview the young people concerning their experiences of different management styles. Any therapeutic community commissioning such research is likely only to commission researchers who are sympathetic to their aims, are aware of the impact of others on the dynamics of the therapeutic relationship, are aware of child safeguarding issues and are able to communicate with young people in a range of mediums. It is also likely they would want to see some acceptance and willingness to abide by an ethical code informed by social work principles. However, it is unlikely that the management researchers' own code of ethics is likely to have covered these issues in sufficient detail. It is not being suggested that social work cannot benefit from alternative disciplinary research or that social work researchers cannot undertake worthwhile research in other disciplinary realms. Such a position would be patently untenable. What is being suggested is that we need to be careful and begin to differentiate as to when social work questions, problems, issues and research strategies require the acceptance of a social work ethical code as opposed to an alternative disciplinary and/or occupational code. This is important, as highlighted at the start of this section, as each code represents a degree of self-interest for that profession and is contextual and temporally situated.

The Ethics of Social Work

England (1986) argues that the choice for social workers is not whether their actions have an ethical dimension or not, but whether this dimension is made explicit and open to question. As Hugman and Smith (1995) rightly point out ethical issues are at the heart of social work.

> Herein lies the crux of the problem, because value statements, being views about what is desirable in society, are highly contentious. They say 'what *ought* to be the case' (Shardlow, 1989: 3), and so open up the potential for disagreement between individuals on grounds of belief and perception (for example of politics, culture or religion). Not only does this mean that an activity like social work will always reflect values, because it is required to intervene in important aspects of everyday life, but that it will often be disputed because the goals of social work may not necessarily be equally acceptable to every member of society. (Hugman and Smith, 1995: 1)

Social work has a long history of engaging with ethical issues and various attempts have been made to classify the principles of social work practice. Probably the most famous and one of the most influential of these has been the work of Biestek, an American Catholic priest who identified 'seven principles of casework' (Biestek, 1961). These principles were:

- Individualisation concerns the recognition of each client as unique.
- The purposeful expression of feeling.
- Controlled emotional involvement.
- Acceptance – unconditional acceptance of the client as a person.
- Non-judgemental attitude.
- User self-determination.
- Confidentiality.

Reflexive Questions

List what you consider as the important values you hold as a social worker. (Keep these safe; you will need them again later in the chapter.)

Compare these values with those identified by Biestek.

What were the major differences between your values and those of Biestek?

Why Do You Think This Was So?

These principles have been surprisingly influential given that they were not initially developed as social work principles. Biestek's emphasis is on the voluntary one-to-one casework relationship of the client and caseworker. In this type of relationship the client voluntarily contacts the agency and relates individually to the caseworker. Many social workers today will find this form of relationship far removed from the complexity of their experience as social workers. Banks (1995) has compared the social work ethical codes propounded by Moffet (1968), Plant (1970), CCETSW (1976), Butrym (1976) and Ragg (1977) and came to the conclusion that they were similar to Biestek's with the Kantian key theme of respect of persons (Downie and Telfer, 1969).

Respect for persons highlights the notion of persons as beings capable of rational thought and able to act with intent. Respect, in this context, represents an 'active sympathy' or love towards another human being. Respect for persons can be criticised for its failure to take into account the position of children who may not have yet reached the stage to be able to act in a rational way although it could be argued they could be considered as 'potential persons'. Similarly how are we to consider those suffering from a chronic mental illness or Alzheimer's – are they to be considered as 'lapsed persons'? The notions of 'potential persons' or 'lapsed persons' may allow both of these groups to be treated to a different and lesser standard than that of other members of society. Respect for persons is seen as important as it reminds us we should treat others as beings that have ends, and are not merely objects or means to an end. The individual is seen as worthy and deserving of our respect simply because he or she is a person. This respect is unconditional.

In the 1980s many social work commentators moved away from the 'list approach'. In particular the 'list approach' was seen as inadequate as these broad principles could be interpreted variously, for example Banks suggests that 'self-determination can mean all things to all people' (1995: 29). Secondly, there is little indication given as to the status of the different principles. Some of these appear to be methods for effective practice (purposeful expression of feelings) whilst others may be regarded as standards for professional practice (confidentiality) and others as general moral principles (self-determination).

Thirdly, how does the code help us to decide which principle to give priority to when two of them come into conflict? What criteria are we to use when deciding whether we should promote a client's self-determination and in so doing reveal a confidential secret?

These principles also came under criticism in the 1970s and 1980s for their focus on the individual casework relationship. During this period a growing awareness amongst social workers that to treat each service user as an individual and to view problems such as poverty, homelessness or crime purely as individual inadequacies resulted in 'blaming the victims' for society's inequalities. A body of writing, which can be loosely termed 'radical social's work', took the view that social work ethics could only be grasped from the perspective of what actually happened in practice. The view that the framing of ethical questions and therefore their answers can only be arrived at from the class positions of those involved can be seen in the work of Bailey and Brake (1975), Corrigan and Leonard (1978) and Simpkin (1979). This body of work was informed by Marxist perspectives and was critical of the notion of 'individualisation' that was viewed as taking the individual out of their social and political context. This is in contradiction to the Marxist approaches which emphasise a materialist framework locating human individuality as being formed by the divisions within society and in particular instanced by divisions of social class.

Banks (1995) does not consider that this literature had a direct impact on the literature of social work values, nor would it have intended to, as one of its key themes is 'praxis' the notion of 'committed action'. This views values, theories and practice as integrally related and it would make no sense in the Marxist approach to separate them. This helped to create a deeper understanding of oppression and to inform later critiques as developed by the feminist and anti-racist movements and has now found its way into mainstream social work values (Ahmad, 1990; Day, 1993; Dominelli and McLeod, 1989; Shah, 1989). Whilst this body of work is often critical of the Marxist-inspired writings their work can be seen to have grown out of, and alongside, the 'radical social work' tradition. A concern for anti-oppressive practices was later incorporated into the list of values produced by the Central Council for Education and Training in Social Work (the forerunner to the GSCC) (CCETSW, 1991b). The rules and requirements for the previous qualifying social work award, the Diploma in Social Work (Dip SW), recognised the competing value positions that social workers were expected to develop including an awareness of structural oppression, the need to understand and counteract stigma and discrimination at both individual and institutional levels, and the need to promote policies that were anti-discriminatory and anti-oppressive (CCETSW, 1991a: 16). (See **Chapter 7** for a fuller discussion.)

Banks (1995) traced the influences of the Kantian, radical and utilitarian perspectives' influence on social work ethics. She then compared 15 national social work codes of ethics and came to the conclusion that there was a significant degree of congruence. There was though a degree of variance between codes as to the degree to which they acknowledged societal and agency contexts. She also identified differences in their impact upon practice and practical guidance over such things as charging fees or user access to records. The congruence focused on shared values underpinning social work and included:

- Respect for the individual person.
- Promotion of user self-determination.
- Promotion of social justice.
- Working for the interests of users. (Banks, 1995: 92)

Reflexive Questions

Looking back at your list of ethical principles, is there any changes you would like to make?

How does your new list compare with Banks' view above?

Can you see any conflicts between the principles?

All four of these concepts are not straightforward and can mean different things to different people, but they do reflect the complexities of social work and begin to identify a value base. The standards for the degree in social work are split into six key roles of which the last one requires qualifying social workers to 'demonstrate professional competence in social work practice' and includes the requirement that social workers are able to 'manage complex ethical issues, dilemmas and conflicts' (Department of Health, 2002a). This is further expanded in the subject benchmark statements for qualifying social work training which cover the nature and historical evolution of social work values; the moral concepts of rights, responsibility, freedom, authority and power; the complex relationship between justice, care and control; the solution of value dilemmas in both inter-personal and professional contexts and conceptual links and conflicts generated by codes held by different professional groups.

The benchmark statement identifies the moral dimensions of social work:

> Social work is a moral activity that requires practitioners to make and implement difficult decisions about human situations that involve the potential for benefit or harm ... Although social work values have been expressed at different times in a variety of ways, at their core they involve showing respect for persons, honouring the diverse and distinctive organisations and communities that make up contemporary society and combating processes that lead to discrimination, marginalisation and social exclusion. (Alcock and Williams, 2000: 2.4)

These statements from the Quality Assurance Agency for Higher Education (QAA), although useful, highlight the difficulties that any attempt to state universal principles are bound to fail. As Mackie (1977: 15) noted 'There are no objective values'. Universal ethical principles cannot help but fail; they cannot refer to a shared general conception of social and moral good – such a view is inherently contestable. Hugman and Smith (1995) point to the work of MacIntyre (1985) to avoid the danger of descending into a spiral of relativism. MacIntyre refers to the eighteenth century Enlightenment for a conception of ethics that saw moral authority and its principles rooted in a particular society at a particular point of time. Ethical principles can thus never be finished but are always open to debate and disagreement and are by their very nature situated and impermanent. However, this does not mean that we cannot begin to identify what is good social work or what are the virtues of a good social work researcher. In particular, MacIntyre argues, in contradiction to the Enlightenment way of thinking, in certain circumstances it is possible to infer 'ought' from 'is'. Thus when we speak of a 'good' farmer, nurse or social worker we have some idea of what a farmer, nurse or social worker ought to be. This notion goes beyond mere 'competency' and brings to the fore the specific virtues of each calling. Husband's (1995) account of the 'morally active practitioner' is helpful here in articulating this struggle to achieve professional ethical guidelines that can never be irreducible to routinised statements.

Given these conditions and an awareness of the limitations of ethical codes Butler (2002) identifies an ethical framework for social work research. His code has been formerly adopted by the Joint University Council's Social Work Education Committee and has also been incorporated into the British Association of Social Workers' *Code of Ethics for Social Work* (BASW, 2002). This model is based on the principles of respect for autonomy, beneficence, non-malevolence, justice and 'scope'. 'Scope' relates to the process for deciding 'to whom and in what circumstances the moral obligation applies' (Butler, 2002: 243).

'Respect for autonomy' refers to the moral obligation to respect the autonomy of each individual in so far as it does not interfere with the autonomy of others. Like respecting persons, it implies the need to treat people as ends in themselves and not simply as means. Respect for autonomy thus begins to identify the need to secure informed consent, preserve confidentiality and to act as to avoid wilful deceit. 'Beneficence' and 'non-malevolence', at their most simple level mean 'doing good' and 'avoiding doing harm'. These terms in the world of medicine can be reduced to a rational calculation as to the costs and benefits to a patient. If the costs are greater than the benefits then the treatment or intervention should not take place. 'Justice', the fourth of the principles, refers to the need to deal fairly in the face of completing claims. As such justice requires the researcher not to further his or her own interests at the expense of others' legitimate interests, for example, in using scarce resources for unprofitable ends, favouring one's own community of interest and disapproval of the moral choices of others.

Butler (2002) cites Gillon (1994: 186) as to how these principles can be combined to identify other moral obligations, for example, the empowerment of others is essentially a combination of beneficence and respect for autonomy whereby the researcher is obliged not only to respect the patient's autonomy but also to enhance it. The application of these moral principles represents the 'scope' or space in which individual researchers can act in accordance with their own moral conscience. This requires the 'morally active researcher' to consider which principle is to take precedence under which circumstance and to be in continuous engagement with their research project.

Social work and the health services have an often-strained relationship but the four principles and 'scope' are very much in line with Banks' (1995) review of other social work codes of ethics and the four key principles she identified underlying these codes. In many ways Banks' four principles could be substituted as social work's four principles with the addition of 'scope'.

Butler has attempted to use these principles to identify a code of ethics for social work and social care research and offers them, 'hesitantly, on the assumption that there exists a body of morally active social work researchers to contest it and develop it' (Butler, 2002: 245). The code of ethics for social work practice is a major step forward for social work research in the UK. Such a code can be seen as an attempt to contribute to the establishment, maintenance and professionalisation of social work researchers. It can also be viewed as providing guidance for practitioners on how to act and as a means to protect research subjects from malpractice or abuse. Codes do though have certain limitations. Homan (1991) recognised that a code implies some means of obligation to behave in a particular way, which in turn implies that the professional group is identifiable and a clear line demarcates members from outsiders. In some occupational groups, maybe medicine, this is achievable but the notion as to who is and who is not a social work researcher is more difficult to police. This becomes even more problematic when we consider the growth of service user involvement in research (as discussed in **Chapter 6**). There is also the question as to what teeth such a code has, especially as at present there is no single regulatory body for social work researchers, although those social work researchers who are also members of the British Association of Social Workers (BASW) could be expelled from the organisation and presumably de-registered by the GSCC. However not all social work researchers will be members of BASW or registered with the GSCC.

Butler (2002), in his criticism of other codes, has already identified an inherent weakness of any code in that it cannot be the final word on ethical matters and will therefore need reviewing and updating. There is also a danger that the code will invite practitioners to find loopholes and to 'play the system' when their research does not fit easily into the code. This might be done around how researchers might justify the use of deception to obtain their data. A final and fundamental concern is that any code will not be able to cover all eventualities. Social life does not fit easily into a commodification of rules of behaviour.

So far this chapter has considered why ethics are important and suggested that if social work research is to have an ethical base it should be linked to social work practice. We have identified a brief history of social work ethics and a proposed code of ethics for social work and social care research. This chapter now discusses a number of the key ethical issues associated with the research process. In particular it will focus on ethical issues before the research commences, during the research and after the research is completed. It would be naïve to suggest that research is a straightforward sequential process whereby one stage leads neatly into the next: as researchers know research can often be a messy disjointed affair. It is though conceptually helpful to divide the research process into a series of stages even if it is acknowledged that it is unlikely to pan out like this in practice.

Ethical Issues Before the Research Commences

Research Governance Framework

Prior to any research being undertaken it will need to be appraised and approved by an ethics or research governance committee. Ethics and/or research governance committees are constituted to regulate research and to quality assure that research proposals and processes are ethical. These committees have been established in universities and cover both staff and student proposals. They are also well established in the NHS but have only in recent times become a serious issue for social services. In order to address this the UK Government introduced the *Research Governance Framework for Health and Social Care* (Department of Health, 2001b). The central purpose of the proposed framework is to ensure 'participants are protected from the risks associated with research' (Department of Health, 2004: 1.1.3).

In order to achieve this the government has established the responsibilities of a research sponsor, who in most cases is also likely to be the research funder, but is not necessarily so. The research sponsor is expected to ensure that the research is worthy of being undertaken and to help research sponsors do this the framework includes a list of key points to examine in relation to any research proposal:

- The research proposal is worthwhile, of high scientific quality and represents good value for money.
- The principal investigator, and other key researchers, have the necessary expertise and experience and have access to the resources needed to conduct the proposed research successfully.
- The arrangements and resources proposed will allow the collection of high quality, accurate data and the systems and resources being proposed are those required to allow appropriate data analysis and data protection.
- Intellectual property rights and their management are appropriately addressed in research contracts or terms of grant awards.
- Organisations and individuals involved in the research all agree the division of responsibilities between them.
- Arrangements are proposed for disseminating the findings.
- All scientific judgements made by the sponsor in relation to the responsibilities set out here are based on independent and expert advice. (Department of Health, 2001b: 3.8.6)

Whilst this is a very helpful list there is a problem as to how, or whether, the requirements of a research sponsor can be enforced by the Department of Health. Also, as Lewis (2002) points out, what would happen where a sponsor could fulfil all the requirements bar one? Would this require the research proposal to be turned down?

The guidance recognises that there are important differences in research in health and social care. Medical research may involve life-threatening conditions requiring very invasive interventions. The threats to individuals in social work research are likely to be of a different order and are more likely to relate to emotional or psychological harm. This may result in differing governance mechanisms, but research governance in both realms should still achieve the same standard. In particular the Research Governance Framework (RGF) focuses on achieving key standards in five domains to ensure good research governance.

- Ethics: ensuring the dignity, rights, safety and well-being of research participants.
- Science: ensuring that the design and methods of research are subject to independent review by relevant researchers.
- Information: ensuring full and free public access to information on research and its findings.

- Health and safety: ensuring at all times the safety of research participants, researchers and other staff.
- Finance: ensuring financial probity and compliance with the law in the conduct of research. (Department of Health, 2004)

The focus of the research framework is on all research involving service users or carers and social care staff for whom directors of social services or social care departments have a responsibility. The RGF is designed to be inclusive and covers all forms of disciplined enquiry that set out to address clearly defined questions by the systematic collection of data, using recognised research methods and techniques including both in-house and externally commissioned research. What is not so clear is whether those activities that public bodies who use most of the same data collection techniques will be classified as research or not. For example, local authorities are expected to collect data for Best Value reviews or Scrutiny Commissions that if carried out by a university would normally be deemed as research (Lewis, 2002).

The UK government now requires all new social care research to have a sponsor who will ensure that the arrangements for the research are in place, accountabilities are documented and clear, independent scientific review has been undertaken and the research team has the necessary resources to deliver the research. For student projects, the academic supervisor, on behalf of the university, will normally undertake the role of the sponsor (Department of Health, 2001b: 3.8).

It will be interesting to see how, if at all, these new arrangements affect the research undertaken in social care and whether the format of external review undertaken by Councils will reflect that undertaken by the NHS which Lewis (2002), amongst others, views as relatively unsympathetic towards qualitative research proposals. Furedi (2002) has also suggested that research ethics committees are more concerned with protective paternalism and act as bureaucratic gatekeepers using ethics as the new managerial ideology. There is also the potential for researchers, engaged in multi-sited studies, having to present their research proposals to a number of such committees each of which may require minor adjustments to the research process which all taken together result in changing the nature of the research. In relation to ethics/research governance committees they have the power and responsibility to veto an application where it is not up to standard. However, it should also be remembered that gaining ethical committee/research governance committee approval does not mean that the researcher can stop thinking ethically. As most researchers realise, although there are clearly ethical dimensions to research design, the difficult ethical issues often arise once the research has gotten underway. For example, how do you know a young person or a person with a learning difficulty or someone with no speech has given their informed consent? This is before we get into issues involving confidentiality, sexual abuse, domestic violence or elder abuse that may arise in the day-to-day research activity. Some of these questions may not have 'right answers' but represent a matter of judgement. Ethical approval within the RGF is conceptualised as a one-off event but research is a process, and as we identified earlier the researcher needs to be continually engaged in reflecting upon the ethical dimensions of their research.

Informed Consent

Reflexive Questions

Imagine you have been invited to take part in a research study. What would you want to know about the study before you agreed to take part?

When would you want to know this and in what format would you want this information?

Under what circumstances might you decline the invitation?

What could the researcher do to alleviate these fears?

A central issue of research ethics that is covered in all research text-books and all ethical research codes is the need for informed consent. Informed consent implies that all human participants should be fully informed about a research project before they agree to take part. This means that all participants should be provided with information about the purpose of the study, what participation is required of them, who the sponsors are and the researchers who make up the research team. This information is also likely to cover what subjects they may be asked, whether participants will be identified or comments attributed to them in the report, whether and if so how participants will be able to find out the results of the study. The participant should also be informed that they have the right to exit the study at any time. This information may also provide details on how to make a complaint and may request a signed declaration from the participant to say they understand what they have been told and that they wish to continue with the research.

Informed consent should also imply that participation is voluntary. This issue is of particular importance when people who have a professional relationship with the research participants carry out the research. Under such circumstances research participants may feel an obligation or gratitude towards the researcher and provide overly positive results. At the other end of the spectrum participants may fear that they will receive a poorer service or that services will be removed if they do not take part.

The notion of informed consent is not without its difficulties, for example it begs the question how informed is informed? There is a balance to be struck here as too much information may curtail involvement unnecessarily or curb spontaneous views. However, as Lewis (2003) comments, there is nothing to be gained from participants who are inadequately prepared. There are potential difficulties for participants in a hierarchical work structure when asked to comment about the organisation's management style unless the participants are convinced that senior management approves of their involvement and their comments will not be personally identifiable. Informed consent becomes particularly important where there is a possibility of adverse consequences for the participant.

Having gained informed consent the researcher needs to decide how they will evidence this. This may be in the form of an oral or written agreement. The difficulty with an oral agreement is that in any dispute it will be one person's word against another's. Personally, written or recorded consent is preferable.

Covert Methods

In relation to informed consent it is important to consider the ethical issue of covert methods and their place within social work research. Covert methods involve the concealment of the researcher within an otherwise acceptable social role (Homan, 1991). Covert research usually involves participant observation, but may also include the use of new technology to 'bug' using video-recorders or tape-recorders. At the start of this chapter we mentioned the example of the 'Tearoom Trade' that is universally condemned as unethical. The question though remains as to whether all social research that uses covert methods is unethical, or, to put it another way, are covert methods ever justifiable?

Homan (1991: 109–13) identifies 13 arguments against the use of covert methods:

- Covert methods flout the principle of informed consent.
- Covert methods help erode personal liberty.
- Covert methods betray trust.
- Covert methods pollute the research environment.
- Covert methods are bad for the reputation of social research.
- Covert methods discriminate against the defenceless and powerless.
- Covert methods may damage the behaviour or interests of subjects.
- Covert methods may become habitual in the everyday life of the person doing the research.
- The habit of deception may spread to other spheres of human interaction.
- Covert methods are invisibly reactive.

- Covert methods are seldom necessary.
- Covert methods have the effect of confining the scope of the research.
- The covert researcher suffers excessive strain in maintaining the cover.

This is a very long and compelling list and suggests that covert methods are at best highly controversial and problematic – if ever justifiable.

Humphries and Martin (2000) cite the case of Davidson (Davidson and Layder, 1994) as a defence of an invasion of privacy. Davidson and Layder studied power and control in the interactions between a sex-worker (Desiree) and her clients. In order to conduct participant observation research Davidson took on a number of roles including acting as the sex-worker's receptionist. Davidson listened into conversations between the 'punters' and Desiree and also discussed the sexual preferences of her punters with her. Davidson defended this covert research strategy by stating that the clients remained totally anonymous to her and that Desiree had willingly offered the information. It was also very clear that the researchers had little time for Desiree's 'punters':

> I have a professional obligation to preserve and protect their anonymity and to ensure they are not harmed by my research, but I feel no qualms about being less than frank with them, and no obligation to allow them to choose whether or not their actions are recorded. (Davidson and Layder, 1994: 215)

Davidson and Layder (1994) then suggest that virtually all social research is exploitative and intrusive to some degree. Social research, and this is also true of social work research, is rarely undertaken at the bequest of either its human subjects nor without due regard to the career of the researcher. This example ably demonstrates that the ownership of information is not always a straightforward manner. It could be argued that the researcher behaved ethically towards Desiree, who participated with informed consent whereas the 'punters' may understandably feel aggrieved that they were duped and that their information was illegally and unethically obtained without their consent. As Punch (1998: 179) observes, 'where you stand will doubtless help to determine not only what you will research but also how you will research it.'

Accompanying covert methods is also the use of deception for which Homan (1991) identifies four types: concealment, misrepresentation, camouflage and the acquisition of confidential documents, which are then used to contact potential respondents.

Homan (1991) argues that where covert methods have been used the researcher has claimed that the end justifies the means. As the above example shows this can be a highly ambiguous and contentious argument. Most research codes prohibit covert research and the deceit inherent in it, although they may contain a rider suggesting that it could be justified in an exceptional case. Butler (2002) provides such a rider when in point eight of the code of ethics for social work and social care research he proposes three conditions before deception or concealment can be used. These conditions are that there should be no alternative feasible strategy, no harm can be foreseen to the research subject and that 'the greater good is self-evidently served'. Any social work researcher considering covert methods would seriously need to consider these ethical dimensions of their research and can also expect a major challenge from their ethics and research governance committee and from their peers if the research enters the public arena.

Anonymity and Confidentiality

Anonymity and confidentiality are sometimes confused in research. Anonymity means that those outside the research team will not know the identity of participants. This may become compromised where participation is arranged through a third party, like an employer, or in case studies where there is structural linkage between samples (Lewis, 2003). In these situations absolute guarantees cannot be given and research respondents should be made aware of this before they agree to participate.

Confidentiality refers to ensuring that the attribution of comments, in reports, presentations or externally published works cannot be linked to individual participants. This includes both direct attribution, where comments are linked to a name or specific role and indirect attribution where a collection of characteristics may make it possible to identify an individual or small group. It needs to be remembered that if the organisation has singleton roles, for example Director or Staff Development Manager, any comments from these roles, or about these roles will be identifiable to those who know the organisation being described. Indirect attribution needs to be carefully scrutinised for if the detail given about an individual or group is too distinctive that individual or group will become identifiable. In order to avoid this situation the researcher may have to restrict the amount of contextual detail that is given.

Ethical Issues During the Research

It is during this phase of the research process that there is the closest interaction between the researcher and the researched. Such interaction inevitably generates ethical issues that may not have been identified in the original research proposal. In social work research this may result in a young person identifying that they have been abused, or are using illegal substances, or planning to break the law. When situations like this occur the researcher needs to remind themselves they are acting as a researcher and not as a social worker (even if they are professionally qualified and registered with the GSCC). In these circumstances they need to be able to refer these issues onto those who can deal with them appropriately having originally identified any limitations to the participant's confidentiality in undertaking the informed consent (see **Chapter 7** for a fuller discussion of research with vulnerable groups). It is also interesting to speculate whether a researcher who is a qualified social worker would be deemed to be in breach of their professional code of conduct if they did nothing in such circumstances?

One key area, which raises significant ethical issues during research, is that of recording data. Focus groups, unstructured interviews and semi-structured interviews are generally tape-recorded. Videotaping is sometimes used, and does have the advantage of being able to record physical gestures, facial expressions and bodily postures but may be intimidating to research participants. Note-taking though, is not generally sufficient for ensuring the accuracy of recording and the actual words spoken, let alone the often-important emphases and silences in question responses. A researcher taking notes is also less able to concentrate fully on the research interview, as they will be concerned about ensuring responses are written down legibly rather than understanding what the respondent is telling them. Where researchers wish to include videotaping or tape-recording they need to ensure that the reason for this is explained and that this is included in the information provided in securing informed consent. This should also include information as to who will listen to the tape, how it will be transcribed, how it will be used, where such data will be stored, for how long and what the procedure is for destroying the tape. Tapes and transcripts should not be labelled in such a way that would compromise anonymity and that identifying information like sampling documents or frames should be stored separately.

Researchers also need to bear in mind that a tape-recorder may appear quite threatening to a participant. To help overcome this the researcher may wish to let the participant hear him/herself before officially starting the interview and place the microphone where the participant can turn it off.

The Right of Respondents to End Their Involvement in the Research

Oliver argues that it is:

> arguably part of the principles of freedom and autonomy inherent in taking part in research that the participants should feel free to withdraw at any time. Even when participants give their informed consent, they cannot necessarily be expected to anticipate their feelings about participation. They cannot anticipate whether they will find the experience enjoyable or stressful. Some parts of the research process

> may prove to be disconcerting, for example in the case of being interviewed about one's personal feelings. It is important that as part of the induction and informed consent process, participants are reassured that they may withdraw from the research at any time. They should not have to give any notice about withdrawal, and they should not have to provide any explanation. (Oliver, 2003: 47)

Oliver highlights the ethical imperative that respondents have the right to withdraw their involvement in the research at any time and without prejudice. This is particularly important as one of the key principles of respondent involvement was that it should be voluntary. This can be difficult for researchers to accept as should all their respondents withdraw they would not be able to complete the research!

There are though a number of reasons why a respondent might withdraw their co-operation. No matter how clear a researcher has thought they have been in informing a participant about the research the participant may not have fully understood what was required of them and it was only once they were answering the questions that it all became clear. Similarly, it is not always possible to plan every aspect of the research and changes of detail need to be renegotiated and this may also include the participant's participation. It may also be that the research raises painful issues for the participant that they are not willing to acknowledge, or are too painful, and they thus decide to withdraw their consent.

Protecting Participants from Harm

Reflexive Questions

Imagine you are a researcher undertaking a study to understand why those suffering domestic violence regularly return to their violent partner before telling anyone. What practical considerations might you need to consider before interviewing those who have been physically assaulted and emotionally abused by their partners?

How might these arrangements differ if you were interviewing the violent partner?

How would you justify this?

In any research study it is important to consider from the outset whether it is likely that taking part may be harmful to individual respondents, and if so to take evasive action or plan contingencies. This may arise directly as in studies on sensitive topics or indirectly when sensitive subjects are raised in relation to otherwise non-sensitive material.

Where subjects are asked to participate in a research study known to cover sensitive material the potential respondent should have been advised of this when their consent was being sought. Lewis (2003) offers the advice that sensitive topics or issues are best addressed through clear and direct questions, so that ambiguity can be avoided and respondents are not drawn unwittingly into areas they would prefer to avoid. The researcher also needs to be sensitive to the respondent's discomfort, and to check whether the participant is willing to carry on. It may on occasions be necessary to stop an interview to allow a respondent to regain their composure before asking them whether they wish to continue or revisit previous answers.

Researchers must also guard against 'mere voyeurism' and be able to clearly justify why certain questions are being asked and how they are essential to the study and contribute to knowledge development. Any subject area is potentially sensitive to someone. This is irrespective of whether the question is about family composition or health status never mind issues like under-age sex, euthanasia or domestic violence.

In the example above you might need to consider the interview venue – somewhere the abused person feels safe – and whether they need to be accompanied by a friend and whether there needs to be a counsellor on-hand should the interviewee become distressed. You will also need to advise the respondent about what types of information can remain confidential and what types of information cannot. This could include information about his or her own abuse but might involve clarifying that any ongoing abuse of a child or vulnerable adult would be reported.

In relation to the abuser you will also need to consider where to interview and how this is likely to impact on the information given. This interview is more likely to take place at a neutral venue. If it is decided to interview in the abuser's home you will need to consider the possibility of being accompanied, for example by a colleague or probation officer, although this is likely to impact upon the quality of the data collected. It is also good practice for researchers to ensure another member of the research team is aware of where they are when undertaking home interviews. Health and safety issues are major considerations in both cases and will, by necessity, impact upon how welcoming and open the interview process is likely to be. The researcher should also consider what their response would be if the abusers should admit further abuse to another partner or request help with their behaviour.

As can be seen the responses to both interviewees are very similar although the degree to which the researcher takes cognisance of the relevant issues varies. Both deserve respect as people, even if the actions of one of them are unacceptable to the researcher. It would be just as bad for a researcher to over identify and sympathise for the abused as it would be to vilify the abuser. Besides suggesting that right and wrong are simplistic categories such views would become reflected in the study distorting the potential for knowledge development. This does not mean that the researcher should not have a view on such issues, but it does mean they should be aware of what it is, be clear to any reader what that view is and to seek to reduce the bias their view results in. This way the reader is able to make their own assessment as to the impact of researchers' views on the results produced.

As mentioned before it is not only sensitive subjects that are likely to bring forth difficult ethical issues – these are also likely to emerge in apparently non-sensitive subject areas. In these cases seemingly 'non-sensitive questions' may raise distressing responses. It is not uncommon for researchers to be faced with the ethical dilemma of being told very sensitive personal material to a question that did not require such a response. One colleague recently was asking a young person about their educational experience only to be told that she had been sexually abused. The young person wanted to share this information and have it recorded before the interview could progress. This can be very difficult for any researcher and particularly for a researcher who is uncomfortable in dealing with such emotional matters.

Reflexive Questions

At this point consider your position as a researcher if you have told the interviewee that everything they tell you is confidential. Are you able to break this confidentiality?

If yes under what circumstances would this be done?

How would this vary depending on the age and maturity of the respondent, or is it inviolate?

This leads onto an important point – the researcher's role is not to be confused with that of adviser, counsellor or social worker. The researcher should avoid giving advice, commenting favourably or unfavourably on a participant's decisions. The researcher, if they are researching a known sensitive subject area should have, as part of their research plan, identified competent professionals who could offer appropriate support and advice should a respondent become distressed. This is not a role they should take on themselves. As part of the research plan for 'non-sensitive' subject areas they should also consider what they should do and whom they could refer participants to should such a situation as the one identified above occur. In such circumstances the researcher may be required to call a halt to the interview and advise the young person that they are duty bound to advise the local child services department. This may seem an obvious action to take, but it does present clear difficulties. To pass on information without the participant's consent is a profound loss of control for the informant. The consequences of passing on information may be hazardous for the participant and raise the likelihood of further harm. This though may depend on a range of other factors like the age of the young person, or where they are currently living and with whom. Certainly the situation of a mature 18 year-old living away at university may be treated differently to an immature 10 year-old living with the perpetrator of the abuse. Even so, as Lewis (2003) acknowledges, this judgement of the degree of risk is likely to involve some subjectivity and may not be one that is shared by those being interviewed.

Protection of Researchers from Harm

It is not only those who are researched who may be at risk of harm but also those who conduct the research. Thought should be given at the start of the research as to the necessary arrangements to minimise any foreseen risk. Risk arises in different ways and may include the travel arrangements to a public place or interviews in private homes.

In public places this may include the funding of appropriate modes of transport and the prior identification of appropriate meeting places at mutually acceptable times. All of this may indicate the need for interviewers to work in pairs either conducting the interview together or one escorting the other to promote safety.

Lewis (2003) reminds us that although fieldwork often takes places in private places it is a public engagement. Others should know where the researcher is and this is likely to offer a degree of protection to the researcher. Risk will always be present no matter how carefully planned and where possible the potential risk factors need to be identified, reduced and managed. If available, background information may help with the assessment of risk and, whilst researchers need to avoid stigmatising assumptions, they should not ignore information and act on it where appropriate.

Researchers may also be at risk due to the nature of the interview or the questions included in the interview. When entering private homes, or unfamiliar buildings to interview, researchers should make themselves aware of an escape route from the home should one be needed. Researchers need to be continually monitoring the emotional level of the interview and be aware that they might need to stop the interview should they feel threatened or worried about their own welfare. It may also be prudent to agree to contact a colleague after the interview not merely for debriefing but to confirm that you are safe and well. Opportunities for debriefing and providing support to researchers need to be built into the process. All these actions may cost money and should be budgeted for in the design stage of the original research budget application.

Ethical Issues After the Data Collection has been Completed

Sharing the Data

The researcher's ethical responsibilities do not end once the data collection has been completed. Respondents may wish to check the accuracy of their information and may request a copy of their questionnaire or tape. Respondents though do not have a right to other respondent's views. Respondent A should not be able to access respondent B's views without B's permission.

Respondents may be sufficiently interested to find out the conclusions of the research. If this is the case they may need to wait until the research is in the public domain in the shape of a published report or article. However, if the original research contract notes that the intellectual rights of the report is the research sponsor's it will not be within the gift of the researcher to share the report without previously gaining the research sponsor's permission. Where possible researchers should advise respondents where they will be able to gain access to the research results and this should have been an issue included in the original ethical approval application.

Publication of the Research Results

For the researcher the end of the research may well be the publication of the results in a book or an article in a peer-reviewed journal. This too is not without its ethical quandaries. Researchers need to discuss publication as part of the original research agreement with the research sponsor and should also have been discussed with informants and any co-operating organisations very early on in the research Homan (1991) suggests that possibly the biggest irritant for a social researcher is the requirement by a research sponsor to approve all forms of publication prior to their submission. This can include not only the final research report, but also any articles or conference presentations derived from the research. Many research contracts now include a clause ensuring that any publications will be seen first by the research sponsor who retains the right to deny the right to publish the material.

In relation to the final research report the research sponsors may not like what has been written about them or their organisation. There are a number of issues here. It is only right that a research sponsor or researcher should be able to see what is written and to be able to correct any factual inaccuracies. Even with the most scrupulous of sponsors there may be difficulties where the sponsor finds it problematical to distinguish between requesting a change because of an inaccuracy as opposed to the findings being inconvenient or not their desired outcome. This is particularly the case when there is disagreement about an interpretation. Often, but not always, a compromise can be reached without jeopardising the integrity of the material or the researcher.

Many researchers will also have faced the situation where the sponsor does not want staff member's responses reported. This might be because the report does not reflect what the manager wants to reflect, the research undermines a favoured policy or practice or the research reflects badly on the manager. This type of situation can be very difficult for a researcher to manage ethically. and raises the thorny question of academic freedom.

> Within the research community the most persistent concern is for the maintenance of academic freedom because it is supposed that the quality and validity of research are dependent upon it. Government investment in research, it is feared, too often has the motive of legitimising intended policy and is too seldom open to findings that would contradict or embarrass policy intentions. Academic freedom is a principle that espouses the discovery of truth without fear or favour. (Homan, 1991: 137)

It is quite possible to substitute any research sponsor or commissioner for the Government in the above quote. The pursuit of academic freedom is undertaken against a background of competition for research contracts – research centres need money and researchers need funding to develop research careers. Under conditions like these there is always a danger that the principle of academic freedom will be compromised.

Authorship

Reflexive Questions

Imagine you have just completed your dissertation or PhD and have written an article of 5,000 words from your work. On showing this to your supervisor they make some changes and suggest where you should publish. At the bottom of their page of comments they request that their name should be added to the work, how does this make you feel?

Do you think is it fair or not?

Authorship can be a tricky ethical matter. Butler suggests that all publications of social work findings 'should properly and in proportion to their contribution, acknowledge the part played by all participants' (2002: 247). There is an immediate difficulty here to define what is meant by 'properly and in proportion to their contribution'. At one level this opens the door to different types of contributions, say around sharing and developing ideas, critiquing work and helping to structure the research. What it does not do is help identify how much of this needs to go on before someone should be considered a joint author. It also raises the difficult issue of how much help a supervisor can provide a dissertation or PhD student before the work is no longer that of the student.

For the student's part, they may feel aggrieved for being put in this difficult position. On the one hand they may not see any valid reason to agree to the request. They are the ones that have completed the research and it is they who have written both the dissertation/PhD and the article. Also, is the supervisor not employed to offer this help and that should surely be reward enough? On the other hand, they may feel they cannot say no as the supervisor is in a hierarchical role to them. On the positive side the supervisor may have helped with the planning of the research, contributed a critical eye to the analysis, read and commented on earlier drafts and supported the student when the going got tough. It certainly would not be easy to refuse such a request. Oliver (2003) suggests that in trying to arbitrate in such a situation we need to consider the different elements of a research article, which includes the academic content, and the writing of the article. In looking back at the example it would appear the student did most of the writing except for some minor proof-reading. The supervisor will automatically have contributed to the research design although this was part of the original dissertation/PhD design. In this type of case it would seem that the major work is that of the student's and that maybe what should happen is a note should be added at the end of the article thanking the supervisor for their support in the completion of the research, which was the basis of the article. This would represent a proper acknowledgement of their efforts on behalf of the student. This example also raises the ethical issue of the exercise of undue influence by the supervisor. It was they who made the request and it is rather difficult to separate this from the issue of power and authority that comes from the role of supervisor. Whilst it is expected that students will listen and act on their supervisor's suggestions it is also expected that supervisors will not act to abuse their position of authority (Oliver, 2003).

Confidentiality and Publication

It is usual for respondents to be given a guarantee, or contractual agreement, that they will not be identified in the report or any subsequent publication. Confidentiality is offered as a condition of informed consent. Pragmatically, confidentiality may not only assist respondents to be more truthful about their views as they cannot be identified but also it may help to guard against a 'halo effect' as there is no purpose in presenting a good impression. One aspect of this, less open to control, is how to ensure neither harm nor embarrassment comes to those who have been researched, or those undertaking research after the report and its findings have become known. Kimmel (1988) reported long-term negative consequences of those involved in the Cambridge Somerville Youth Study 30 years after the original project. Punch (1998) also argues that it is important to avoid the sense of betrayal felt by some interviewees when they find their information has been used for purposes that were unclear or unknown about at the start of the process. Once published it is very difficult to predict what uses one's research might be put to. Thus whilst it is possible to give a guarantee of confidentiality there is no guarantee that once published the work will not incur negative effects on those involved nor that respondents will feel angry about how their data has been used.

Summary

Ethics are integral both to social work practice and to social work research. This chapter began by identifying some of the abuses in the name of research demonstrating why ethical research is so important. Having established the need for ethical research the chapter then argued that if there were to be a claim for a distinct social work research ethics this would be related to the ethical nature of social work and social workers. There then followed a discussion of social work ethics starting from the work of Biestek. From this it was established that social work ethics were situational and contextual, and although not fixed, there was a congruence of an international social work ethical position around the principles of respect for persons, honouring diversity and challenging injustice. This was then related to the nature of social work research ethics and Butler's (2003) BASW code for social work and social care research. It was also shown how this code relates to the principles of respect for autonomy, beneficence, non-malevolence and 'scope'. From this discussion it was concluded that, although very important, ethical research codes only provide a framework and there is still a need for the 'morally active social work researcher'.

The chapter then moved on to discuss ethical issues before the research commenced, during the research and after the research was completed. In particular we focused on informed consent, anonymity, confidentiality, covert methods and research governance and ethics committees before the research commenced: data collection and storage, rights of respondents and the protection of respondents and researchers from harm during the research. Following the research we discussed the storing of data, publication of results and the particular issue of confidentiality and publication. All this underlined the importance for social work researchers to be continuously engaged with the ethics of their research, seeking to maintain academic freedom and credibility whist steering a defensible course between respondents, research sponsors and the wider social work and academic communities. It is worth highlighting the first point of the ethical code for social work and social care research here:

> At all stages of the research process, from inception, resourcing, design, investigation and dissemination, social work and social care researchers have a duty to maintain an active personal and disciplinary awareness and to take practical and moral responsibility for their work. (Butler, 2002: 245)

Suggested Reading

Banks, S. (2006) *Ethics and Values in Social Work* (3rd edn). Basingstoke: Macmillan. A well-written, easily readable guide to the major ethical issues and dilemmas facing social workers.

Butler, I. (2002) 'A code of ethics for social work and social care research', *British Journal of Social Work*, 32(2): 239–48. Identifies the code and reasoning behind a separate code of ethics for social work and social care. Now also included as a section in the BASW Code of Ethics.

Homan, R. (1991) *The Ethics of Social Research*. Harlow: Longmans. The standard text referred to by most British social science researchers.

Oliver, P. (2003) *The Student's Guide to Research Ethics*. Maidenhead: Open University Press. A good introductory text on key ethical issues for social science students.

5 Evidence-based Practice – Panacea or Pretence?

This chapter will identify how research has sought to influence practice. In particular the chapter will examine the importance of the growth of 'evidence-based practice' and how it has spread from the field of medicine to many other areas of human endeavour including social work. The chapter will consider how 'evidence-based practice' has been differently interpreted within medicine and the fields of social work and probation. The chapter also provides a critique of evidence-based practice and the different ways it can be used to support practice and policy.

Reflexive Questions

What do you think evidence-based practice means?

Can you write down your definition?

Can you think of any research evidence that you have used to inform your practice?

Where did you get your research from and how did you know the research was credible?

Evidence-based practice has been one of the success stories of the 1990s. When I began writing this chapter I put 'evidence-based practice' into the Google search engine and identified 3,490,000 references. Sheldon et al. (2005) remind us that evidence-based practice (EBP) is not a new phenomena in social work and point to the work of Joseph Rowntree and Mary Richmond along with the President of the American National Association of Social Workers who in 1931 stated:

> I appeal to you, measure, evaluate, estimate, appraise your results in some form, in any terms that rest on anything beyond faith, assertion and the 'illustrative case'. Let us do this for ourselves before some less knowledgeable and less

> gentle body takes us by the shoulders and pushes us into the street. (Cabot, 1931: 6, quoted in Sheldon et al., 2005: 12)

In recent years EBP has rapidly become a global phenomenon with practice manuals, methodological networks, research centres, journals, newsletters, toolkits, software packages, websites and e-mail discussion groups (Trinder, 2000a). EBP has also been trumpeted in various government publications including *Modernising Social Services* (Department of Health, 1998) and the national *Framework for the Assessment of Children in Need and their Families* (Department of Health, 1999a). The last of these publications describes EBP as a key principle of assessment that should be grounded in evidence-based knowledge (1999a: 10). It is also seen as a key component for good organisational resource strategies that are characterised as committed to evidence-based practice (Topss, 1999) and in the guidance for qualifying social work courses (Department of Health, 2002a). EBP has spread from an approach championed by medicine to cover most health-related fields including dentistry, public health, nursing and physiotherapy and to more distant fields including management, education, probation and social work. EBP would appear to promise clear and simple answers to difficult problems and offer certainty in an otherwise uncertain world. This book's concern is with social work, but we cannot begin to understand EBP without first returning to medicine.

Supporters of evidence-based medicine claim that it developed in order to bridge the gap between research and practice. In medicine this gap is potentially, quite literally, life threatening. For example, Antman et al. (1992) demonstrated that the majority of contemporary text-books recommended treatments for myocardial infection that were of proven worthlessness.

Sackett et al. (1977: 71) are generally accepted as the starting point for evidence-based medicine with a widely accepted and quoted definition:

> the conscious, explicit and judicious use of current best evidence in making decisions about the care of individual patients, based on skills which allow the doctor to evaluate both personal experience and external evidence in a systematic and objective way.

Conscientious, explicit and judicious are all terms open to challenge and interpretation. Conscientiousness refers to being diligent and taking great care and at another level acting in response to one's conscience. We would all probably agree that professionals like doctors and social workers should act diligently and in accordance with their conscience, so long as their conscience is reflective of our own ethical system. For example, do we wish a doctor to allow a terminally ill patient to die with dignity even if this may mean shortening their lifespan? Questions like this are more difficult to answer and are likely to be dependent upon our own ethical and moral code. Explicit refers to something being precisely and clearly stated whilst judicious implies proceeding from good judgement. Evidence in social work has not reached the position whereby everything can be precisely or clearly stated, nor in working with people is it ever likely to be a question of mathematical or causal certainty. Working with people still requires a passion for humanity, a belief in human potential and a willingness and commitment to go beyond normal expectations. It is a truism to say we want professionals to act with good judgement, but who is to decide what 'good judgement' is and what is good enough 'evidence' to base that judgement on. Inherent in the medical version of EBP is the notion that the doctor knows best.

Reynolds (2000) notes that another important feature of evidence-based medicine is that it distinguishes between research that is of direct clinical relevance and research that is not. If research findings have no relevance to clinicians they are of marginal interest to EBP.

The Cochrane Collaboration, an international not-for-profit organisation, has supported and promoted the development of evidence-based healthcare through providing up-to-date information about the effectiveness of healthcare interventions. The Cochrane Collaboration produces and disseminates systematic reviews of healthcare interventions to help people make informed decisions about healthcare.

These reviews are based on the hierarchy of evidence stated below with the most credible and trustworthy at the top and the least credible at the bottom.

1. Several systematic reviews of randomised control trials or meta-analyses.
2. Systematic reviews of randomised controlled trials.
3. Randomised control trials.
4. Quasi-experimental trials.
5. Case control and cohort studies.
6. Expert consensus.
7. Individual opinion. (Becker and Bryman, 2004)

The Cochrane Collaboration's hierarchy of knowledge can thus be seen to favour quantitative research at the expense of qualitative. The collaboration currently has 12 methodology sub-groups of which only one is related to qualitative methodologies and even that is shared with the Campbell Collaboration. The Campbell Collaboration (C2) was founded in 2000 to help make well-informed decisions about effectiveness in the social, behavioural and educational arenas (www.campbellcollaboration.org.uk, accessed Dec. 2004).

MacDonald (2003) has written a SCIE report on systematic reviews for improving social work. In this she identifies the following four key features of systematic reviews:

- Systematic reviews are reviews that have been conducted in ways that minimise the chances of system bias and error.
- Systematic reviews are characterised by explicitness and transparency. Protocols are an essential feature of a systematic review.

- The best systematic reviews are produced by teams comprising users, practitioners and researchers.
- Systematic reviews do not remove the need for judgements. (MacDonald, 2003: 3)

Evidence-based Probation

Within the social welfare field EBP has been most enthusiastically taken up by the probation service. Probation officers work with offenders in the community and prisons whilst also being an integral part of Youth Offending Teams. In England the probation service is located in the Home Office whilst in Scotland they are positioned with social work services. Probation officers work within the controlled environment of prisons and in the community. Probation also has a history of being well served by government-funded research.

In the 1970s probation, and work with offenders, was best summed up by the phrase 'nothing works'. The 'nothing works' era in British probation is usually highlighted by the IMPACT study (Folkard et al., 1976). This famous study allocated probation officers at random to 'standard' or 'intensive' caseloads to investigate whether the results would be better when probation officers had more time to work with their clients. The overall result was that those who received the 'intensive' probation, the experimental group, were slightly more likely to re-offend than those who only received the normal probation programme. This study suggests that getting more probation officer input made things slightly worse than having less probation officer input. However, it should be noted that no advice was given as to what 'intensive' meant and there is no means of comparing what actually happened with the two groups.

This rather depressing message of 'nothing works' slowly began to be challenged. By the 1990s the probation service began to rediscover its faith in its own methods as positive results were found in learning-theory-based methods of practical work with offenders. The then Chief Probation Inspector, Sir Graham Smith, saw the principles and methods of EBP and the new 'what works' as offering a foundation for the development of a new and valued role for the Probation Service. This later led to what Raynor (2003) has described as:

> Recent history's most spectacular example of wholesale conversion to evidence-based practice can be found in the National Probation service of England and Wales. (Raynor, 2003: 334)

To back up this assertion Raynor notes that the first 'what works' conference occurred in 1991, then the launch of the Effective Practice Initiative in 1995 along with the publication of McGuire's edited collection of 'what works' conference papers (McGuire, 1995) and the 'Underdown Report' on effective supervision (Underdown, 1998). These were then followed up by the launch of 'what works' pathfinder projects and the Joint Accreditation Panel in 1999. Correctional Services Accreditation Panels were then established to quality assure accredited practice programmes which were likely to reduce offending as they had an adequate theoretical base, evidence of effectiveness, good training arrangements, good materials and a commitment to monitoring quality of delivery and outcomes. The promotion of 'what works' has engendered an optimism that appropriately implemented accredited interventions can lead to a reduction in re-offending and a subsequent reduction in the prison population. Reducing offending is not only a moral and technical issue but also a political one as the accreditation panel has been aware of conflicting drivers – on the one hand the need to be certain about effectiveness and on the other the need to accredit for wider implementation across the country.

Merrington and Stanley (2000, 2004) have reviewed the evidence on the effectiveness of 'what works' interventions in probation. Most of the studies included in the evaluation were based on cognitive behavioural principles. In assessing the impact of the research they noted that the results were often less than those suggested by the research. This disparity is potentially explainable by both the complexity in the intervention and its evaluation. Programmes are difficult to evaluate and compare and as yet there is no clear 'winner' among the range of available interventions. They do not suggest that evidence-based probation is finished but argue for more targeted research to help identify which programmes are the most promising and which are not.

Underdown (1998) and Chapman and Hough (1998) also point out that the introduction of accredited programmes is not sufficient in itself to address offending behaviour – this still requires the exercise of good case management skills. This additional factor creates problems in deciding upon effectiveness, as good case management will ensure that the offender is admitted to the right programme when they are receptive whilst supporting and reinforcing their learning during and after the programme. All of this creates difficulties in attributing cause and effect as the research has yet to distinguish between outcomes in response to good case management and those due to offender motivation (Merrington and Stanley, 2004).

Overall, Raynor (2003) accepts that this revolution within the probation service has not been without its critics. He accepts that EBP has yet to sufficiently accommodate issues of diversity, tending to focus on majority as opposed to minority groups, and has not been well served by its managerial arrangements. He does though note that it is not dependent on psychologists or cognitive behavioural approaches and that a return to some golden bygone age of 'nostalgic illusion' would be a mistake. Instead, he argues that:

> the way forward for probation services is more likely to be found by broadening and extending the evidence-based approach than by abandoning it. If the National Probation Service wants to pursue the reforming mission shared by most of its staff by providing alternatives to more punitive and less constructive sentences, it is not enough to show that an ever-increasing prison population is not cost-effective. It is also necessary to *demonstrate* (not simply to claim) that community penalties can provide greater tangible public benefit, more reparation to the community, greater opportunities for offenders to change their behaviour, and in the long run a reduction in offending. (Raynor, 2003: 342)

Evidence-based Social Work

The development of EBP has been somewhat more contentious in social work. This may be because probation is a much more focused service whose primary role is to reduce offending whilst social work has a much wider service user base and broader societal remit. This is not to say that EBP has not made inroads into social work practice:

> It is perfectly possible for the good hearted, well meaning, reasonably clever, appropriately qualified, hard-working staff, employing the most promising contemporary approaches available to them, to make no difference at all to (or even on occasion to worsen) the condition of those whom they seek to assist. (Sheldon and Chivers, 2000)

Sheldon, one of the major proponents of evidence-based social care practice in England reminds us that as social workers it is not enough just to have good intentions. The actions of social workers can either improve or worsen a situation, or possibly even worse, have no impact at all.

EBP rightly reminds us of the importance of outcomes. Too often, it is argued, social work has become overly preoccupied with processes at the expense of the task. In the current climate of performance indicators and star systems, outcomes are often seen as all that matters and it should be remembered that the best, and most enduring, outcomes will be achieved through effective processes.

'What Works'?

In recent years there has been a plethora of publications about 'what works' e.g. on early years (MacDonald and Roberts, 1995), child protection (MacDonald and Winkley, 1999) or leaving care (Stein, 1997). All of these publications identify that generally successful interventions are characterised by being well-organised, structured, focused, time limited and contractual. The production of such research-based practice suggests that interventions can be replicated and workers trained to deliver these interventions. The most extreme case of this view was experienced by the author at a conference for Chief Executives to mark the introduction of Youth Offending Teams. At this conference a Chief Executive offered his vision of the future for youth justice workers. In this future workers would only be allowed to intervene using previously accredited evidence-based programmes. Having assessed an offender the worker would then match the nature of the offence and the offender's behaviour to an accredited programme and implement the intervention. The programme, as it had previously been accredited would work and if it did not, he would be asking the worker what they had done wrong!

Reflexive Questions

How do you feel about the example of the Chief Executive?

How would you feel about being held to account for your interventions in a similar way?

What reasons might you use to suggest why some intervention might not work?

What reasons might you suggest why some intervention does work?

How would you prove to a colleague that it was your intervention that had been successful?

EBP contains an inherently positive feel. It challenges bureaucratic logic – the logic that proclaims that things are alright as they have always been done this way and politics, which promotes the art of the possible rather than the logical (Nutley and Webb, 2000). In fact, it would appear almost heretical for a research book to be critical about EBP. However I would like to raise some questions about the current overly medically dominated characterisation of EBP. To begin with I would like to give the reader a health warning, no evidence of no effectiveness is not the same thing as evidence of no effectiveness. Social workers should not necessarily stop doing something just because there is no evidence that it is effective. For example, we should not stop treating service users or colleagues with respect even if there is no evidence to suggest that doing so will produce a better outcome.

I would like to critically examine some of the claims for evidence-based social work. In particular I would like to focus on the nature of social work, the lack of evidence and epistemological issues.

EBP and the Nature of Social Work

The first point to be made here is that social work is not medicine. Social work is not able to control the environment in which practitioners work, nor is there a prescribed level of intervention doses that will cure social work problems. The evaluation of social interventions presents problems not encountered in other, more clinical settings. Social work's service users are not passive recipients of social work practice. Each service user will have their own unique history and understandings, their own unique trajectory of personal and interpersonal problems many of which may not lend themselves easily to monitoring or evaluation. These difficulties and issues are constructed in social interaction with family, friends, work colleagues, acquaintances and the like which remain outside the control of the social worker. Service users are quite likely to have their own view of what is wrong and what needs to happen to put it right. Problems are rarely unidimensional, but rather multidimensional impacting on theirs and others' lives. Issues like housing, employment, support networks and caring duties can all affect a carer's mental health.

Social work encounters are unlikely to be straightforward. Social workers have to work with complexity and contradiction with individual, families, group's or community's 'messy' problems and in so doing a monocausal explanation or assessment is unlikely to lead to an effective intervention. Social workers work in a social, moral and political context and cannot disengage from the world in which they work and inhabit.

The extreme form of the above argument suggests that social work represents an art form. EBP challenges this view suggesting that practice should be scientific and objective. This model represents social work as a rational scientific exercise. The rational actor model assumes that social workers will act on the available evidence in a 'clinical' way regarding social workers as 'information processors' operating within a closed model of decision-making (Webb, 2001). The position taken by this author is that social work is both an art and a science inextricably entwined in the personal and political. In fact it is the very complexity of social work that demands social workers engage their hearts and their minds to develop a creative and accountable practice that is research-minded and uses the best knowledge available. If social workers are to avoid the danger of pathologising individuals whose problems are largely the result of wider social factors, and at the same time avoid being rendered impotent by the range of competing theories and explanations, they need the skills to be able to read and assess competing alternatives' knowledge sources.

Critique of the Scientific Basis of Evidence-based Practice

Trinder (2000b) and Webb (2001) both argue that EBP is predicated on a particular model of science as positivistic and behaviourist. Random controlled trials, the 'gold standard' of this model (see **Chapter 3**), allow researchers to postulate causal connections giving them an edge over other forms of social research. RCTs potentially allow us to control for bias arising from extraneous variables and through randomisation they provide a degree of high internal validity. There are though difficulties in considering RCTs for social work. MacDonald (1999), a keen supporter of RCTs and evidence-based practice, acknowledges the realities of social work do not always make it possible to randomly allocate service users, the desirability of large samples are not always practical and the high costs of RCTs make them less attractive to social work researchers. Potentially of greater concern to social work is their lack of explanatory power given that there is often a loose relationship between our understandings of a social problem, for example domestic violence, and our responses to it. If we were measuring the effectiveness of support groups to reduce the re-occurrence of domestic violence how would we know whether this was due to the model of domestic violence management, the charisma of the support group leader, the companionship of the other women or even changes to the violent partner that were most important in creating a change? The study would not be able to tell us what forms of support group work with which victims of domestic violence, or why and under what conditions they work.

This limited explanatory ability leads to a limited purchase on being able to match intervention to service user. The ability to match effective interventions with service users is of critical importance to a practically oriented profession like social work, which demands that its practitioners engage with service users to address their problems and issues. On top of this there is also a classic aggregate net-effect problem (Trinder, 2000b) whereby the positive effects of an intervention for some are cancelled out by the negative effects for others. Overall this results in a mean score, which is neither positively or negatively associated with an intervention that works (or does not). RCTs can identify which interventions are more broadly effective than others, or even compare to where there has been no intervention at all. They are though limited in their ability to provide practitioners with a clear toolkit of effective interventions.

The EBP cookbook of RCTs, meta-analysis, experimental and quasi-experimental designs are all located within an, often unacknowledged, value system and philosophical version of what counts as knowledge and how that knowledge can be known.

> Evidence-based practice proposes a particular version of rational inference on the part of decision makers. It assumes there exist relatable criteria of inferential evidence based on objectively verdical or optimal modes of information processing. In other words it creates a picture of social workers engaged in an *epistemic process* of sorting and prioritising information and using this to optimise practice to its best effect. (Webb, 2001: 63, italics in original)

EBP assumes that social workers, as rational agents, will be able to draw out the logical consequences of EBP findings. In this they will be able to apply logical principles about the likelihood of actions achieving desired ends that are informed by a behavioural probability calculus. This ignores the context in which social work decisions are often made. Social work decisions are often dependent on decisions outside a rational framework. Social work decisions may be governed by legal statute, agency imperatives, the politics of inter agency relationships and managerially led initiatives especially where these are likely to impact upon performance measurement agendas. Social workers are not free agents – they operate within a limited and circumscribed rationality. EBP's failure to acknowledge this position seriously weakens its usefulness and appropriateness for social work.

Lack of Evidence

Within evidence-based practice there is an assumption that we have sufficient knowledge to direct practice. Social work has yet to reach this point. MacDonald et al. (1992) examined 95 research studies, many of which were American, from over 50 journals, which attempted to measure social work effectiveness through the use of experimental, quasi-experimental, pre-experimental and client opinion studies. The experimental studies included randomly allocated equivalent-group designs that provided, potentially, the most persuasive evidence of effectiveness. However such an approach is difficult, both ethically and practically, to achieve in social work. There are major ethical issues in, for example, elder abuse if you decide not to intervene in a group of cases for the sole purpose of measuring the effectiveness of some new intervention or treatment. Quasi-experimental designs regularly have no random allocation and no pre-intervention of matching of intervention and non-intervention groups. Client opinion studies seek the views of service users of the effectiveness of the services they receive and whilst this is an important indicator of effectiveness it is only one indicator and needs corroboration from other evidential sources. Overall, these studies covered some 18 client groups and 16 different methods of intervention. Of these studies it would be assumed that the majority would cover high profile areas of social work, but this was not the case as only 13 attempted to evaluate child protection work. This dearth of empirical evidence in child protection was also mirrored in other key aspects of social work practice including work with older people and mental health.

In another review of child protection work, Oates and Bross (1995) identified a range of inclusion criteria including sample size of more than five subjects, have a method of comparison and at least 15 per cent of participants in the sample were known to have been physically abused. Over a 10 year period the authors were only able to uncover 25 papers that met the selection criteria whilst none of these were concerned with routine child protection social work provision.

Trinder (2000b) notes that even in the United States, where empirical research work is at its most advanced, the outcomes of experimental and quasi-experimental work has achieved little. Muluccio's (1998) research into child welfare outcomes in the United States came to the conclusion that findings across a wide range of studies were inconclusive whilst many studies were bedevilled by methodological problems including issues of inadequate control, underspecified interventions and outcome variables. In response to this, Parker (1998) has developed a hierarchy for social work knowledge defining knowledge as tentative, indicative or conclusive. Tentative knowledge is where the evidence suggests we should pause and take stock. Indicative suggests that we should proceed with caution and conclusive represents the highest level of scientific reliability and validity. For Parker the bulk of social work's knowledge base remains within the indicative range with only a small amount being conclusive.

This makes it particularly difficult for social work to slavishly pursue an EBP approach when the evidence is currently insufficiently developed to support practice. There is also a debate to be had as to whether social work can, or ever will, reach such a point. This is not to say that we should not pay due cognisance to what research evidence is available, but it does mean that we are not currently, or may never be, in the position to be totally guided by research evidence. For the evidence-based researcher this represents a call to arms in that if we would only provide the resources they would be able to develop the evidence base to answer practitioner questions. This perspective purely sees evidence-based research as a technical exercise in solving problems. One of the difficulties of EBP has been its colonisation by the positivist's rational scientific view that inhibits and constrains its potential to answer researcher and practitioner questions alike. As Trinder (2000c: 236) comments:

> Evidence-based practice tends to fit solutions to problems into its own world view, providing more and better information, further refinement of methodological criteria or incorporating consumer perspectives into an evidence-based practice framework. The result is that there are major outstanding issues that the rational scientific model is ill equipped to handle, and would appear unlikely to resolve.

From Trinder's perspective evidence-based practice is thus too limited a perspective in which to place all our faith for research to support practice.

Evidence-based Practice and the Pragmatists

Trinder (2000b) also makes the interesting observation that EBP in social work has not only been promoted by the 'what works' camp but also by those wishing for research to have a much greater impact on practice whom she describes as pragmatists. These two camps are both currently claiming EBP. For the pragmatist evidence-based practice primarily means practice informed by research. Research in this instance does not only mean RCTs but also includes other types of research including practice wisdom and service user views. Amongst these pragmatists exist the Research in Practice (RiP and RiPfA), Making Research Count (MRC) initiatives and the Social Care Institute of Excellence with its governmental mandate to promote effective social care research.

RiP is a funded partnership between the Dartington Hall Trust, Association of Directors of Social Services, the University of Sheffield, local authorities, and private and voluntary social care organisations. RiP originally described its aim as promoting 'evidence-based practice and policy in child welfare services' (Trinder, 2000b: 147) but the strapline has since been changed to 'supporting evidence-informed practice with children and families' (www.rip.org.uk, accessed 7 Aug. 2005). This shift from evidence-based to evidence-informed is highly significant and symbolises a distancing from the pure medical evidence-based model to a more inclusive research-based model. RiP maintains an 'evidence bank' that contains a range of research information derived from a diversity of methodologies that has been assessed as useful for practitioners. Uniquely RiP targets councillors and senior managers in its mission to disseminate research findings in addition to practitioners and front-line managers. Alongside these initiatives RiP also produces research reviews, research and policy updates, CD ROMs and audiotapes of research, a regular newsletter and jointly, with Making Research Count (MRC), the Quality Protects Research Briefings. RiP has been traditionally associated with child and family social work but following a merger with the Centre for Evidence Based Social Services launched an adult arm, Research in Practice for Adults (RiPfA).

MRC is a different type of national initiative, describing itself as a national initiative with local presence. MRC is currently centred on nine universities each with a lead social work professor. Each university base has a number of paying partners drawn from local authorities, independent social care and health agencies. There are some agencies that are members of both RiP and MRC. Similarly to RiP, MRC is concerned with improving dissemination and implementation of research findings but describes its relation to evidence-based practice slightly differently. MRC (G Man) based at the University of Salford, one of the newer hubs of the partnership, states that part of its core principles include:

- Recognition that evidence-based practice brings together different sources of knowledge: research knowledge, social worker knowledge, practitioner expertise and agency findings resulting from reviews, auditing and monitoring.
- Awareness of the importance of organisational arrangements for an effective evidence-based strategy and building capacity in organisations to develop research-mindedness amongst practitioners. (www.chssc.salford.ac.uk/scswr/making_research_couint/index.shtml, accessed 11 Nov. 2005)

SCIE has already been identified in **Chapter 1**, and is an independent company and charity partly funded by government with the express intention of developing and promoting knowledge about good practice in social care. SCIE provides a wide range of free reports, knowledge reviews, research briefings, position papers, practice guides and resource guides and the UK's most extensive social care database, social care online (www.scie-socialcareonline.org.uk, accessed 7 Mar. 2005).

All three of these organisations, along with organisations like the Joseph Rowntree Foundation (www.jrf.org.uk), What Works for Children? (www.whatworksforchildren.org.uk) and the Mental Health Foundation (www.mentalhealth.org.uk) are committed to the dissemination of research although the evidence on social workers' willingness to read the social work literature is less encouraging:

> the shorter and less obviously informed by research the piece, the more likely it is to be read by social workers. Community Care is regularly read by nearly half (47%) of professionals but only a fifth (19%) frequently turn to specialist books and fewer still (15%) use national guidelines often. The most depressing finding was that 29% never read a specialist book and a fifth were willing to say they never open the guidance which effectively brings to life the Children Act 1989. (Bullock et al., 1998: 67)

Although the research above focused on children and family teams it is hard to imagine that the situation with adult teams is any different. In my own experience of social work *Community Care* has always been very popular, but what this high readership rate does not tell us is whether the workers read Community Care for its articles, or just for the job adverts. Nor does it take into account that for many workers this is a free publication delivered to social work offices. In response to this concern about the dissemination of research both RiP and MRC have developed experimental projects to identify more effective and creative ways of providing research evidence for social workers.

Clearly, organisations like RiP, RiP*f*A, MRC and SCIE are committed to ensuring that research makes an impact on practice. There is a compelling argument that social work practice should be informed by research to ensure the optimisation of effective outcomes. However, as we have already seen, social work is not identical to the biological basis of medicine, nor does it have a sufficiently sized evidence base and importantly, nor would RCTs necessarily provide the answers to social work questions. At this point you may wish to consider the question that even if there was clear research evidence would you, or your social work colleagues implement research findings that seemed to contradict what you believed to be the case? This is a question we will return to later.

For the pragmatist there is an acceptance that if research can be helpful to practice it is useful, irrespective of the research paradigm or method which produced it. Even Sheldon (Trinder, 2000b: 148) has commented that if social work research were to be subject to the same inclusion criteria as used by the Cochrane Collaboration very little of it would be accepted.

Evidence-based Practice and the Future

It would be all too easy to suggest that evidence-based practice has become a political slogan and has no place within social work. To support such a view would be to undermine the potential contribution that evidence-based practice offers social work to become a more research-informed practice. Evidence-based practice has rightly focused on outcomes, stressed the need for reducing bias in research, promoted internal validity of research and sought to encourage service user perspectives. Importantly the whole debate around evidence-based research has raised fundamental questions as to what should constitute evidence and to how should research inform practice.

> It is important that all evidence be treated with caution. The methodological rigour of evidence-based practice conveys a sense of certainty and authority. As we have seen, however, absolute authority is rarely to be found. There are major areas where evidence is lacking, questions about the extent to which evidence can be trusted (meta-analysis), questions about the applicability of evidence in real-life cases and concerns about the narrowness of evidence and narrowness of outcomes. In some areas certainty is more founded, whilst in other areas, beyond the biological, the search for certainty poses considerable danger in inherently complex and uncertain worlds. (Trinder, 2000c: 237)

EBP has also brought welcome attention to the part played by research in informing practice although it has on occasions suggested simplistic solutions to complex problems. Initiatives like RiP, RiPfA and MRC are to be welcomed finding favour with government and social work agencies alike to ensure research evidence makes an impact upon practice.

EBP has though tended to deal with the complexity of the human condition and the political processes of research as if they were technical problems to be eliminated and overcome, suggesting greater certainty than the model deserves or can deliver. As we have seen the model, and its dependency on RCTs, does not transpose well to social work. This is not to say that RCTs are inappropriate in social work, but to say there are limited opportunities to develop RCTs and that the knowledge they generate is not necessarily the most useful to social workers nor should we necessarily privilege this method of generating knowledge above all others. The definition of evidence needs to be expanded to go beyond RCTs and the claims of EBP need to be reduced to mirror their ability to deliver and inform social work practice. It is for this reason that this chapter was titled evidence-based practice – panacea or pretence. Research can help to inform practice but it is not a panacea, it does not stop the social worker having to make judgements and it is also not pretence as research can help social workers to be aware of which interventions or which issues need to be considered in which context. It does though overemphasise the ability of research to direct social work practice and it is probably more accurate to talk of research-informed practice.

Summary

EBP is likely to continue to be an important aspect of UK health and social care policies. This chapter has sought to define EBP and identify how it has developed in probation and healthcare and how it has been transposed onto social work and social care. This transposition has been seen to be wanting when we examined the nature of social work and the narrowness of the definition of evidence. This narrow definition results in an unhelpful hierarchy of knowledge whereby RCTs are seen as the gold standard and stand at the top of the pinnacle. If the healthcare criterion of evidence were accepted it would result in a quantitative social work evidence base neglecting or denigrating the value of qualitative research. The chapter has also highlighted how social work 'pragmatists' have embraced this interest in evidence for practice to raise the profile of research and to promote the potential of research as a means of informing practice. For social work there needs to be a widening of the definition of evidence and at the same time a more realistic claim made for the possibilities of EBP as one of the key determinants informing practice and not merely the only one.

Reflexive Questions

Having read this chapter what do you think should be considered as evidence in social work practice?

How can we know whether this evidence is credible and trustworthy?

What sources of evidence do you look for to inform your practice?

What level of evidence would be needed for you to change a tried and trusted practice?

Is it feasible or desirable to have evidence-based social work?

Suggested Reading

MacDonald, G. (1999) 'Social work and its evaluation: a methodological dilemma', in F. Williams, J. Popay and A. Oakley (eds) *Welfare Research: A Critical Review*. London, UCL Press.

Sheldon, B. and Chivers, R. (2000) *Evidence-based Social Care: A Study of Prospects and Problems*. Lyme Regis: Russell House Publishing. The first chapter of this book entertainingly highlights key issues for social work in relation to evidence-based practice.

Trinder, L. and Reynolds, S (eds) (2000) *Evidence-based Practice: A Critical Appraisal*. Oxford: Blackwell Science. This is an edited collection of articles looking at evidence-based practice in a range of subject disciplines and suggests that the further you get away from the controlled medical ward the more difficult it is to achieve the ideal model of evidence-based practice.

6 Service Users and Research – the Next Frontier

In this chapter we explore the involvement of service users in improving social work practice and in particular focus on the increasing participation of service users in social work research. Before discussing how to involve service users we ask why should service users be involved in research? Having done this we discuss the methodological, ethical and practical issues and how these may be addressed. This chapter also seeks to contribute to the developing debate as to what are the benefits, costs and limitations of involving service users in research? In what aspects of the research process can service users be involved and how might service user involvement add value to the research process?

Reflexive Questions

Before going on much further I would like you to consider what you understand by the term service user?

The term service user is used in this text to refer to those who need and use services provided by social workers and social care workers. Statutory agencies, voluntary bodies or private agencies may provide these services either voluntarily or non-voluntarily. This is not to imply a dualism between those who receive and those who supply social services. Most of us, whether social worker or social researcher will use social services either directly, for example, we may have mental health needs or a child with a disability, or indirectly through, for example, the provision of residential or domiciliary care for a relative. The labels researcher, social worker and service user are not exclusive. Service users, like social workers or social work researchers, are not a homogeneous group. Although it is useful to use the shorthand of service users we should not slip into the position of thinking that all service users think alike. This is not to say that they will not agree on certain issues, but it serves to remind us that we should not make assumptions about what others may think.

The Growth of the Service User Movement in Social Work

Social work has a proud tradition of seeking service user views that goes back to the seminal *The Client Speaks* (Mayer and Timms, 1970). Mayer and Timms systematically recorded the views of social work clients, the term we used to use when indicating a service user, identifying what they felt about social workers, what they saw as helpful and what they did not.

Reflexive Questions

If you are (have been or were to be) a social work service user what qualities would you want from your social worker?

In 1996 the National Institute for Social Work reported on a study on standards for working with service users and carers commissioned by the Department of Health. This report began by stating:

> It is easy to summarise what people who use services and carers value in their contacts with social service workers: they value courtesy and respect, being treated as equals, as individuals, and as people who make their own decisions: they value workers who are experienced, well informed and reliable, able to explain things clearly and without condescension, and who 'really listen': and they value workers who are able to act effectively and make practical things happen. (Harding and Beresford, 1996: 1)

In other words service users and carers respected those workers who respected them, workers who were experienced, able to listen and able to make things happen.

The push for a service user mandate comes from both the consumerist tradition of the 1990s and the democratic tradition of developing participation in order to improve the quality and effectiveness of services. Service user involvement has become embedded in organisational requirements, policy and legislation. The involvement of service users and carers have been central themes in the modernisation agenda of New Labour which put service users at the heart of social care (Department of Health, 1998, 2000a) and the national health service (Department of Health, 2000b). This has been further emphasised in the UK Government's national service frameworks for mental health, older people and children (Department of Health, 1999b, 2001c; Department of Health and Department for Education and Skills, 2004) along with the *Valuing People* strategy for people with learning difficulties (Department of Health, 2001d).

This growing movement has expanded to all aspects of social work. Service users and carers are represented on the General Social Care Council (Hasler, 2003), have been integral to the development of the qualifying social work degree (Barnes, 2002), the approval of social work courses at pre- and post-qualifying (Department of Health, 2002a; GSCC, 2005) and are involved in the planning and delivery of social work education at both pre- and post-qualifying levels (Beresford, 1994; Citizens as Trainers et al., 2004; Molyneux and Irvine, 2004).

Service users and carers identify their aspirations for social workers of the future (Reform Focus Groups, 2002) as:

- The need for social workers to understand what a person's life is really like, and not to make assumptions and judgements about what they think the person needs.
- The importance of the quality of relationships that the social worker has with the service user.

The first of these statements begins to hint at why service user research and service user involvement in social work education has become so influential. The quote highlights the importance of understanding someone else's life from their perspective, not from our own standpoint position, and the importance of the relationship between service user and social worker. As we noted earlier it is important for social work researchers to be aware of their history and how this impacts upon their own understanding of social work issues. This focus on understanding the world from the service user perspective ably leads us to consider the current position of service users in research.

The Benefits of Service Users' Involvement in Research

This section could also have been titled 'Why bother about service user involvement in research?'

> **Reflexive Questions**
>
> Before we begin to explore the nature and some of the claims made for involving service users in research you might want to write down what you perceive as the benefits for involving service users in research.

Involve, whose remit is to promote public involvement in NHS, public health and social care research have identified the following reasons for service user involvement in research (Hanley et al., 2004: 2–4):

- People who use services will be able to offer a different perspective. This point reminds us that although we may be an expert in the matter under research it does not mean we know everything. Without the service user perspective it could be argued that any picture of service effectiveness would be incomplete. Knowledge about service users can be deemed to be incomplete if it does not include the knowledge that service users have of themselves.
- People who use services can help to ensure that the issues that are identified and prioritised are important to them. Service users or carers can identify research topics that would not necessarily be identified by researchers.
- Public involvement can help to ensure that money and resources aren't wasted on research that has little or no relevance. As previously noted social work is an applied discipline and it is important for social work research to address the needs of service users, but also to identify interventions that service users are willing to use and are effective. It would be of little benefit to identify a wonderful new intervention if it was not needed or that nobody was willing to use.
- People who use services can help to ensure that research doesn't just measure outcomes that are identified and considered important by professionals. It cannot be assumed that researchers can identify all the outcomes that are important to service users. After all service users are ultimately the end users of research and understandably are important stakeholders in the effectiveness of the research.
- People who use services can help with the recruitment of their peers for research projects. Service users may understandably be suspicious of the motives of researchers and as such be unwilling to become involved in the research. Working with a voluntary organisation or self-help group can often assist with recruitment of peer researchers especially with hard to reach groups who by definition are less likely to be willing to come forward.
- People who use services can help access other people who are often marginalised, such as people from black and minority ethnic communities. This is the other side of the last point. Being able to work with a marginalised service user as a researcher is likely to provide greater opportunities to be able to identify other potential service user respondents from the same community of interest.

- People who use services can help to disseminate the results of research and work to ensure that changes are implemented. Service users are often keen that the results of the research are effectively disseminated. This may take a number of forms in that many voluntary organisations and user-controlled organisations have their own websites, newsletters and/or magazines that will publicise research summaries. It is also possible that service users can help present research results at conferences. If prepared properly, service users can be a very powerful voice for publicising the key messages from the research (Kirby, 2004).
- Involvement in research can help empower service users. Involvement in research can help to raise self-esteem, develop new skills and help improve employment prospects especially if it involves training.
- The involvement of the public is becoming an increasing political priority.

The Reasons Given why Researchers do not Involve Service Users

Looking at the above it is difficult to see why researchers do not involve service users, but this is evidently the case and we need to look at their reasons in more detail.

> **Reflexive Questions**
>
> Before we discuss the reasons given by researchers you should write down at least three reasons why you believe researchers would not want to involve service users in the research process. Do you think these reasons are justified or not?

Lack of 'Representativeness'

This concern is regularly raised in the literature concerning service user involvement in policy, practice or research. This concern asks how can we ensure that those service users included in the research are typical of all other service users. It is patently unreasonable to assume that one or two people can be representative of all people who use similar services. As we have already identified, the term service users does not signify a homogeneous group and there is no agreed meaning of

'representativeness' just as there is no 'one' service user voice (Molyneux and Irvine, 2004). The Association of Directors of Social Services (Jones, 1995) drew up practice guidance indicating that service users and carers contribute perspectives based on their experience not on their representativeness. Involve also highlights the importance of the service user perspective as opposed to trying to ensure some mechanistic representativeness:

> In essence it is not reasonable to expect one or two people to be representative of all people who use similar services. But then it is not reasonable to expect one doctor to be representative of all doctors either. It might be helpful to reframe this in terms of thinking about seeking **perspectives** rather than representativeness. If you want a range of perspectives, involve a range of people. (Hanley et al., 2004: 5, bold in original)

Service users, especially if they are recruited from democratically constituted groups may be representative, or even typical users but their legitimacy arises from their personal experience. This personal experience is likely to be similar to others using similar services in similar circumstances. Equally, researchers are not a homogeneous group – you just have to read their differing views in the academic journals or in earlier parts of this book. Researchers have diverse views and experiences – they have no greater claim to representativeness than service users.

'The Usual Suspects'

A special case of representativeness is the issue of the 'usual suspects'. This view suggests that it is always the same people who get asked to represent service user views. Attached to the non-representativeness of this approach is often an assumed bias or cynicism that organisations, or researchers, are hand-picking those people who are most likely to agree with them or represent the 'acceptable' face of that service user group. At the extreme this may just be another form of tokenism where there is an illusion of participation but no real change in the locus of power or even willingness to consider change.

However, this is to miss the point. In requesting service user participation in, for example, a research advisory group, you will want to ensure that you involve those people who are able and willing to participate in meetings. They will also need to be able to present their views to a variety of individuals with a range of professional expertise including researchers and research sponsors. These representatives may not be 'typical' but will be able to offer insights from their perspective and may also be able to access other service users' views. Service users may need support in developing their confidence and skills in presenting their arguments whilst chairs of such groups need to be inclusive in the way they structure their meetings.

'The Research can't be Objective if Service Users are Involved'

The above statement presupposes that researchers are objective and neutral but this is not the case. Researchers like service users will want to become involved in research that excites and interests them. People who use services, including researchers, can be supported to avoid becoming overly emotionally attached and biased towards a research study. Keeping service users involved will also ensure that the research findings and any recommendations will remain relevant to service users.

It is possible that the research may become personally upsetting for service users and consideration needs to be given to their personal support. This is an important aspect of the ethical considerations in relation to a research project (see **Chapter 4**). Such support should obviously be considered when researching sensitive subjects – for example, the impact of child sexual abuse or compulsory mental health admissions to hospital – but may also be required for less sensitive subjects. All human beings are different and involvement in the research process may awaken, or trigger, a previous unhappy experience or incident.

'It'll be too Expensive and Time Consuming'

Involving service users and carers will cost more money and take more time (Kirby, 2004; McLaughlin, 2005; Smith et al., 2002). To involve service users or carers effectively is not resource neutral. If service users or carers are to be used as co-researchers they will need to be trained for this role. The True project (Lockey et al., 2004) found that service user research training was essential and led to actual involvement in research and a desire to be involved in more research. The initial costs in terms of time and resources did result in research involvement and insured the investment was not wasted. The training also had enormous value for the participants' personal development, self-esteem and self-confidence. The benefits of service user participation are more often identified in terms of personal development than organisational change (Carr, 2004) and this is one of the key challenges for such research to demonstrate positive outcomes as well as empowering processes. Researchers and research sponsors need to include these extra costs within bids. To avoid doing this is to undermine the quality and potential usefulness of the research output.

'Service Users will have Unrealistic Expectations of Research and its Implementation'

This argument suggests that service users will underestimate the amount of time and effort required whilst at the same time overestimating the impact the research will make on changing services. This is not necessarily a problem if researchers, service users and research sponsors can be clear from the outset about the length of the research project, who will be involved, the nature of the research outputs and the potential for these to be put into practice. What is important here is that researchers and research sponsors should not promise anything that is not within their gift to deliver.

So far this chapter has identified the mandate for service user involvement in research. In so doing we have identified what can be seen as the major advantages of such an approach and some of the reasons given by researchers or research sponsors for not including service user involvement in the process. Hopefully, this has included the reasons you have identified as well. If this is not the case you may wish to consider how your reasons for not involving service users could be addressed. The chapter will now consider the philosophical underpinning of service user involvement in research and to look at different levels of service user involvement, including service user led research.

The Philosophical Underpinning of Service User Research

When considering service user research it is important to consider the assumptions behind involving service users in the research process. Building on the work of Christensen and Prout (2002) with young people it is possible to suggest a typology of four different models. These four models view service users either as objects, subjects, social actors or active participants in the research process. The first of these perspectives perceives service users as objects to be acted on, measured and researched. This is the traditional quantitative model of research whereby the research is undertaken by a researcher who would determine a baseline, undertake their intervention and then measure the changes to the situation, attributing any changes to the nature of the intervention.

The second viewpoint perceives the service user as subject and in so doing brings them into the forefront of the research process. This represents a much more service user centred approach, but potentially has the shortcoming of being tempered by researcher views as to the ability or maturity of the service user. This could be particularly important when considering the position of children and young people, people with a mental illness, people with learning disabilities or those suffering with dementia. Clark (2004) has shown that it is possible to begin with notions of competency when researching the experience of children under five and the importance of viewing young children as 'beings not becomings' (Qvortrup et al., 1994: 2 quoted in Clark, 2004: 142). Such a view can be translated to other marginalised groups who are viewed as not being fully competent.

The third approach views service users as social actors who have the ability to act on, change and be changed by, the world they live in. In this perspective service users are seen as autonomous individuals, and critically, not part of another system whether this be family, work or school.

The fourth perspective is a special case of the last approach whereby service users are seen as active participants in the research process (Christensen and Prout, 2002). In this perspective service users are not seen merely as social actors but as co-researchers collaborating and participating in the research process. This begs two important questions – one concerns the level of involvement of service users and the other as to which, if not all, parts of the research process service users can become involved in. The next section deals with the first of these two questions.

Levels of Service User Involvement in Social Research

There is a critical difference between going through the empty ritual of service user participation in research and service users having sufficient power to impact upon the outcome of the research. Arnstein (1971) has famously identified a ladder of citizen participation, which also transfers to service user research. This ladder consists of eight rungs: manipulation, therapy, informing, consultation, placation, partnership, delegated power and citizen control. The first two of these rungs, manipulation and therapy, are identified as non-participation whilst the next three rungs are seen as mere tokenism with only partnership, delegated power and citizen control viewed as citizen power. A simpler approach by Hanley et al. (2004) identifies three points on the same continuum; consultation, collaboration and service user control. Hanley et al.'s continuum is explored next.

In contrast to Arnstein, Hanley et al. (2004) present a positive view of participation and ignore the possibility that service user involvement in research may be illusory or non-participatory. For example, consider the position of a service user who agrees to be a member of a research steering group. This would appear to be offering a degree of service user involvement, but if the steering group is always held at times when the service user cannot attend, no thought is given to venue or the need for alternative caring arrangements, if the language of the steering group is so specialist or jargonistic that it is exclusive and excluding, it would be totally inappropriate to suggest that this was real participation. In reality this would appear to be non-participation dressed up in the guise of service user involvement.

Consultation

Consultation is integral to modern UK society where we have the notion of the 'listening bank' and satisfaction surveys for everything from the purchase of holidays and toasters to cars. However, asking people about their satisfaction with services is notoriously unreliable. Compliance, unequal power relationships between those asking the questions and those answering them and a lack of understanding of alternative possibilities will all potentially act to influence those being questioned towards a positive response, unless the appraisal instruments are quite sophisticated (Gutek, 1978). Hanley et al. (2004: 8) suggest that:

> When you consult people who use services about research, you ask them for their views to inform your decision-making. For example, you might hold one-off meetings with people who use services to ask them for their views on a research proposal. You will not necessarily adopt those people's views, but you may be influenced by them.

One of the major difficulties with consulting people is that when asking their opinion there is no guarantee for them that you will act on their views although they may influence you. One of consultation's greatest strengths is its imprecise definition that generates what McLaughlin et al. (2004) describe as a 'useful ambiguity' allowing consultation to mean different things to different people.

Practice Example

A local authority and a primary care trust want to know whether deaf and hard-of-hearing service users are satisfied with the quality of services on offer. Together they arrange a meeting for Wednesday at 3 p.m. at a local deaf club. At the meeting only three people turn up and the research officer assumes people must be happy with the services otherwise more people would have turned up.

Looking at the above example do you believe the research officer was correct in their evaluation of the situation? If not, what issues might they need to address to improve the situation? How would you have done it?

I hope you would have first considered that deaf and hard-of-hearing people are not a homogeneous group; deaf people do not necessarily go where hard-of-hearing people go and young deaf people do not necessarily attend adult deaf people's haunts. The choice of venue may at first appear sympathetic to deaf people but this is not necessarily the case – it may not be where the majority of deaf people go, never mind those who are hard-of-hearing. The researcher should have first found out more about their target group to identify where they congregate so as to maximise access. They could have talked to some members of the deaf and hard-of-hearing community to identify where such places are and which is the best day(s) and time(s) to arrange a meeting. They could also have worked with this small group of service users to examine how best to present the issues in a format and language that is accessible and understandable to the service user group. They will need to consider such issues as transport and caring responsibilities. This all needs doing prior to considering issues such as interpreters and communication requirements. Also, and most importantly as McLaughlin et al. (2004) demonstrate you may also need to address what consultation means to certain groups, for example, deaf people who are not used to being consulted with and therefore do not know what is expected of them in such a situation.

Consultation is the lowest level of service user involvement as identified by Hanley et al. (2004). Consultation can be seen to be a simple process, which if you have never involved service users in research before can be a 'safe' place to start. This is particularly the case, as consultation does not commit those consulting to act on any particular suggestion. The locus of power resides firmly with those undertaking the consultation, not those being consulted. Many people who use services find it frustrating to be asked their opinion without any commitment from those asking to act on that opinion. It is unsurprising those communities who are regularly subject to consultations complain of 'consultationitus' or 'consultation overload'. Beresford and Croft (1993) claim that there are compelling reasons why certain powerless groups are regularly asked their views on issues without any tangible outcomes to the process. 'Consultation fatigue' can result in ill health for those individuals or organisations that are regularly targeted without clear outcomes. One other disadvantage of consultation is that you may miss out on service user's ideas as the responses are constrained by the researcher's agenda. In such situations consultation can become a pseudonym for social control and non-participation.

Collaboration

Collaboration occupies the middle ground between consultation and service user control. Collaboration implies a degree of ongoing service user involvement. For example this might include service users being active members of the research steering committee, collaborating on the research design, acting as researchers in analysing and interpreting data, contributing to the writing of the report and undertaking an active part in the publication and promotion of the results. Collaboration can mean any or all of these things. Collaboration exists in the grey area between consultation and service user control, and as such it represents a broad spectrum of participation. It is also important to remember that one person's assessment of a situation as collaborative may not be shared with those who are being collaborated with.

Collaboration increases the likelihood that any outcome measures or assessment criteria will be relevant to the research participants. Again collaboration provides for the possibility that those service users in the project will help access other service users. As such the collaborators can act as informal research sponsors helping with the recruitment and informed consent of potential research subjects. This can be particularly important with hard to reach groups. Moreover, once the data has been collected, collaborative research service users can help with the contextualisation, interpretation and understanding of the data. Collaborative research projects are more likely to result in service users feeling greater ownership of the results and being more likely both to publicise the results and to take onboard any changes required as a result of the research findings.

On the debit side collaboration can be costly in terms of time and resources without any guarantee of a better outcome. Collaboration is likely to result in extra resources for service user time, service user travel costs and other expenses. On top of these costs additional resources may be required, for example in the training of service users, producing plain English versions of research papers and providing for the support and development needs of the service users.

Collaboration, unlike consultation, requires a different mindset from the lead researcher. The researcher must be willing to share ownership of the project and be actively seen to take into account others' suggestions and to behave in a collegial way. For some professional researchers it may be very difficult to accept shared power. Such an approach also demands that researchers may require an expanded skill set to include facilitation and negotiation skills.

Service User Control

User-controlled research is still in its infancy and remains a contested arena. Broadly speaking user-controlled research is located in those research projects where power resides, not with the professional researchers, but with the service users who are responsible for the project and subsequent decision making. This does not necessarily mean that service users undertake every stage of the research by themselves, or that there is no place for 'professional' researchers within this process. The key issue here revolves around the issue of power. Who has the power to make the decisions? As opposed to the consultative and collaborative position service user-controlled research locates decision making clearly with service users. Thus if professional researchers are employed to help with the research they are working to the service users' wishes and desires. One example of user-controlled research involved the Wiltshire and Swindon User's Network who undertook a leading role in designing and undertaking research on the implementation of Direct Payments for disabled people as part of a Best Value Review (Evans and Carmichael, 2002). This style of research is often preferred, and may be seen as the only legitimate research from the viewpoint of some service user groups.

The disability movement has been instrumental in the development of this type of research. The movement sees traditional research as furthering the careers of able-bodied researchers at the expense of disabled people that led to Finklestein (1985) famously stating 'no participation without representation' (quoted in Barnes and Mercer, 1997: 6). Research was viewed as 'ripping-off' disabled participants, by simply using them to obtain information without any involvement in the construction of the research agenda or what was seen as legitimate to research. The roots of this critique of social research on disability can be traced back to at least the 1960s and the case of the disabled residents in Le Court, Cheshire Home (Barnes and Mercer, 1997). The disabled residents asked expert researchers from the Tavistock Institute to support their struggle to take greater control over their everyday lives. This was a three year funded research study that quickly alienated the residents. The 'unbiased social scientists' followed their own agenda and their report rejected the residents' complaints and recommended a re-working of traditional practices even though they categorised institutional life as a 'living death' (Miller and Gwynne, 1972). The residents felt betrayed by the report and labelled the academic researchers as 'parasites' (Hunt, 1981). Several of the disabled people in the Le Court protest later helped form the Union of Physically Impaired Against Segregation (UPIAS) in 1975. UPIAS became central to the development of the social model of disability critiquing experts and professionals who claimed to speak on behalf of disabled people, but who in UPIAS' opinion only pursued their own interests:

> We as a Union are not interested in descriptions of how awful it is to be disabled. What we are interested in is the ways of changing our conditions of life, and thus overcoming the disabilities which are imposed on top of our physical impairments by the way this society is organised to exclude us. (UPIAS, 1976)

Similar to consultation and collaboration, user control is both a point on a continuum and a range of potential options orbiting this point. For example, it is unclear whether user control refers only to the research process and whether it also includes those whose initial idea the research was and who wrote the research bid.

The advantage of user-controlled research is that it is more likely to address questions that are pertinent to service users. It is also possible that it will focus on issues not considered by researchers and can often reveal evidence that would otherwise be missed. As service users or their organisation(s) own the research there will be a high commitment to making an impact upon policy and practice. Being involved in a user-controlled research project can be an empowering experience for those whose previous experience has been concerned with being the research objects of others. Hearing the voices, concerns and recommendations of service users is also likely to impact upon professionals who are concerned with ensuring the quality of service provision.

User-controlled research is not without its disadvantages. User-controlled research requires researchers to hand over power and the control of a project to those who use services. Some researchers, research sponsors and research funders may find this difficult or unacceptable. The research itself may not be viewed as independent as those who have most to gain are those undertaking the research. There is though a difference between a rigorously implemented research design, analysis and evaluation, and biased research. Bias is not a problem solely restricted to user-controlled research and the same techniques used to reduce bias in traditional research can be implemented with user-controlled research.

We have considered four different levels of service user involvement in research: non-participative, consultation, collaboration and user-controlled. All except the first of these may be appropriate in different research circumstances. Different levels of involvement will depend upon the lead researcher, the research topic, the resources available, the research method and the funding body. The involvement of service users in research is not just about having an extra person at the steering group or someone different undertaking interviews. It is also about issues of democracy, social justice, quality of communication, providing opportunities for different contributions and for other contributions to be valued. Involvement is a complex and multifaceted activity where in the same research service users may be required to be consulted, collaborated with or even, required to take the lead:

> The involvement of people who use services, when done well, positively affects not just **what** is done in research, but also the **way** it is done. (Hanley et al., 2004: 12, bold in original)

It should also be noted that service user research is not restricted to people with physical disabilities but can also include any service user group. In principle it should be possible to involve any service user group in research in an intellectually honest manner. The research process does rely on intellectual skills and will therefore be less open to young people, those suffering an episode of mental illness, those with learning disabilities or older people with cognitive impairment, but this does not mean these groups do not have a contribution to make about their lives and their experiences of living those lives. Nor does it mean they will not be able to make an effective contribution although alternative research tools and techniques may require development.

Having considered the different levels of service user involvement in research this chapter now considers some of the practical issues often raised in relation to service user research, in particular, we focus on meaningful involvement, rewards for service users, recruitment, training and the position of children and young people as co-researchers.

Meaningful Involvement

As noted earlier, when researchers seek to involve service users in research, this participation must be more than merely tokenistic. The meaningful involvement of service users requires careful thought and planning to ensure an effective contribution to the research. If participation is poorly implemented it is likely to have negative consequences for the research and for service users who will be less likely to trust researchers in the future and may become cynical about the research process. Poorly implemented participative research not only runs the risk of having negative consequences for the research in hand, but may contaminate the field for research in the future.

Kirby (2004: 10–11) identifies a range of questions to be asked about the meaningful involvement of children and young people. These questions, with minor adaptations are just as suitable for all service users:

- Is service user involvement planned from the beginning?
- Are service users involved in deciding how they want to be involved? Is participation always voluntary?
- Will they be supported to get involved in ways that suit their needs, abilities, interests, access needs and availability?
- Will all the information about the research be shared, so they are able to make informed opinions and decisions?
- Will their views be genuinely listened to, and influence decisions along with the views of other stakeholders?
- Will they be treated as equals? How will you demonstrate respect for their contribution?
- How will they personally benefit from being involved?
- Will disabled members have the opportunity to contribute equally?

In addition to this set of questions thought should also be given as to how service user's contributions will be acknowledged in the research report and any future publications arising from the research. This may be as a co-author, member of the research team, steering group member or whatever. It is also possible that because of the nature of the study, for example in cases of research into sexually transmitted diseases or criminal behaviour, service users may not want to be acknowledged by their full name but by their first name, pseudonym or a nickname. Some service users may also see themselves as championing certain issues, and naïvely believe they are aware of the consequences of 'going public'. It would appear patronising not to allow service users to 'go public' but service users should be reminded that once the research is in the public domain there is no control over how, or by whom, it will be used. It is thus important to be clear about this from the start of the research project and to discuss these possibilities with service users including identifying an agreed publication strategy.

Fair Return for Participation

A guiding principle of involving service users in research should be that there should be no exploitation of service users by researchers. This represents a development of beneficence and non-malevolence discussed earlier. When service users, either as respondents or as co-researchers, become involved in research they may contribute large amounts of time and energy as well as providing access to their networks, language and culture. This is time, knowledge and experience that could have been spent in pursuing a job, domestic work, education or contributing to their own or their family's survival. It is thus important that service users should be recompensed for this time and effort.

Reflexive Questions

You might like to think about how you would recompense service users who were acting as co-researchers. Are there any circumstances in which you would not consider money as an appropriate reward?

Having done this, can you consider what you see as the advantages and the disadvantages to recompensing service user researchers?

Before beginning to discuss recompense it is important to be aware that under the research governance framework (Department of Health, 2001b) all individuals engaged in research which falls within its remit require an 'honorary contract'. This is irrespective of whether the service user is paid or is a volunteer. This requirement seeks to ensure that sponsoring organisations take overall responsibility for the research, that they can meet their obligations to ensure the health and safety of all those involved in the research and cover any liability. If service users are given an 'honorarium' payment they technically become employees, albeit casual workers and health and safety and employment legislation come into force. Recent advice from the Inland Revenue (personal letter) suggests that the Inland Revenue does not consider small amounts paid to cover expenses of service users in the research process, as either respondents or co-researchers, is likely to attract attention as the amounts are unlikely to place the recipients within taxable bands.

The first point to note is that recompense does not necessarily imply cash. The researcher may provide gift vouchers to be redeemed at a local shopping centre or the procurement of a specialist piece of equipment or membership of a local gym as potential rewards. The research may also provide financial support to a service user organisation as an alternative to giving recompense directly to the co-researcher. However, this type of arrangement will not replace any income foregone and may not motivate individual service users.

A co-researcher, like any other researcher, feels valued when they are recompensed for their efforts. Recompense is also likely to increase motivation and can act as a stimulus for the completion of the less exciting research tasks. It is also likely that if service user researchers are being paid not only will they take themselves more seriously but also professionals and other organisations will take them more seriously.

On the negative side once service users are paid there is a change in the relationship that becomes more contractual and potentially controlling. There has now been an exchange of money for services. Paying service users may also impact negatively with service users losing some, or all of their, state benefits and with children and young people it may result in inappropriate pressures for them to act as employees. There is also the potential danger that some people might just join the research to be paid rather than because of any special interest in research or the research topic. Bringing money into the equation potentially identifies a new arena for dispute concerning the need to treat co-researchers with transparent fairness and equity.

However you look at it bringing money into the equation will impact upon the research, for good or ill. What is crucial is that researchers should discuss recompense at the beginning of the research so that service users who consider being involved in research can take this into account in their decision.

The Limits of Service User Knowledge Production?

This section has argued for the inclusion of service users in the research process. However, much of the research studies in relation to service user involvement included in this chapter focus more on processes than outcomes. The process of involving service users can be beneficial to service users but it does not necessarily mean a better research product. Questions are yet to be answered as to how far service users can be involved in all aspects of research. In particular the more advanced forms of qualitative and quantitative research are likely to be beyond what can be taught in a short research methods course for service users. It also begs the question that if service users receive research training to a level that they are able to engage with the more advanced statistical analysis or critical discourse analysis do they become more researcher than service user?

However, it is just as disingenuous not to engage service users where they could benefit a research study as it is to believe they can deliver something they cannot. Lead researchers need to be aspirational in relation to service user researchers whilst still being grounded in what is reasonably possible.

Children and Young People

This section seeks to highlight some of the key issues involved when collaborating or working with young service users as co-researchers. Up to now the chapter has dealt with service users as a totality – in this section we focus on just one of the potential service user groups to act as an exemplar of this type of research. It should also be noted that children and young people might be members of more than one service user group. Children and young people may also be physically disabled, suffer from mental health problems, be asylum seekers or have a combination, or all, of these identifying characteristics.

> There is no right way to involve children, just as there is no ideal technique to use in research with children. You may want to involve children in every aspect of your research, or to consult them about specific areas. (Save the Children, 2004: 42)

As this chapter has demonstrated there is a great value in participatory research. However there will be research topics and contexts where a more traditional approach will be more suited. Controversial topics like child sexual abuse or violence against children may not be acceptable by powerful adults or ethics committees as suitable for research by children and young people.

Some of the issues highlighted in this section are developments of those already discussed in relation to service users generally and may also be appropriate to other vulnerable groups. In particular I would like to highlight the mandate for young service users' involvement, the advantages offered by this participatory approach, the employment position of young people, informed consent and safeguarding young service users.

Mandate for Involvement

The mandate for children and young people's involvement comes from both the children's rights agenda and the government's policy statements. The United Nations Convention on the Rights of the Child (United Nations, 1989) sets out the benchmarks for the rights of children everywhere. However, it should be noted that the UK Government has not adopted the convention as part of its legislative framework and there are conspicuous non-signatories especially the USA and Somalia (Johns, 2003). However, those countries that are signatories are assessed at least once every five years on the extent to which their childcare and legal systems conform to the requirements of the convention. Article 12 is of particular note as it promotes a child's right to participate in decision making by being informed, involved and consulted on all matters that impact upon their lives As Roberts (2004) notes this means that the research agenda in researching with service users is also a participation agenda.

A number of the government publications supporting service user involvement have already been identified in this chapter. In relation to children and young people it is also worth noting the publication of *Listening, Hearing and Responding* (Department of Health, 2002b) includes an action plan for the involvement of children and young people. Government bodies like the Children and Young Person's Unit have claimed that:

> The Government wants children and young people to have more opportunities to get involved in the design and evaluation of policies and services that affect them or which they may use. (Children and Young People's Unit, 2001: 2)

Both the children's rights movement and government policy promote the involvement of young service users.

The Advantages and Disadvantages of Young Service Users as Co-researcher

It is one thing to argue that young service users should be more involved in the design and evaluation of services and policies, but it is another to identify what are the advantages and disadvantages of such an approach.

Reflexive Questions

Think about a research topic that you feel would be of significance to children and young people, for example bullying or divorce. Now identify what you feel are the major benefits of involving peer researchers to investigate your research topic.

Now think of any disadvantages or costs that you can see to involving young peer researchers in your research topic.

There are a number of benefits claimed for involving people as co-researchers as summarised by Kirby (2004). These include the opportunity to enhance the range and quality of data. Young people speak a common language and as a consequence will be able to frame questions and interpret responses more accurately than an adult researcher. Young people will raise issues with another young person that they would not raise with an adult. The work of Butler et al. (2002) on children's involvement in divorce suggests that young people will often turn to friends, or those who have undergone a similar experience, to find out the implications for them of their parent's divorce.

In much of the world the position of children has changed dramatically in recent years. HIV/AIDS, bullying, conflict and other factors make it difficult for adults to understand the reality of children's lives. As Save the Children (2004: 13) comment: 'Adult researchers may have less insight into the daily lives of children than they think they have'. If this is the case then children and young people need to be more involved and included in research.

Young researchers' enthusiasm may act as a motivator to the whole group whilst young people and children may approach old problems in new ways Collaborating with young service users, as researchers, will also ensure the ongoing learning and development of service provider organisations by confirming effective programmes and appropriate services.

Coupled to this point it could be argued that organisations that promote the empowerment and enhancement of the position of children and young people in society should also reflect this commitment in the way they commission research. To do this by definition would mean the promotion of service users as research sponsors as well as service users as co-researchers or peer researchers.

It is also true that young people presenting their findings can have a greater impact with audiences as opposed to an adult researcher. Alongside these reasons I would also add that young people might benefit from learning new skills that increase their confidence and self-esteem whilst at the same time enhancing their employability. This rationale does not mean that it is always appropriate, or the best way to proceed, to include young service users as researchers. Involving young service users as researchers is not without its complexities and difficulties.

In trying to involve young service users as researchers, there can be difficulties in recruitment, the research is likely to take longer, cost more and there may be certain research activities that are less attractive or inappropriate for young people.

Recruiting young people as researchers can be problematical although an existing youth group or organisation could be approached. But, if you want to work with a group of socially excluded young people they are, by definition, difficult to recruit. The very act of recruitment can become a major task and time intensive activity in its own right.

Researching with young service users is likely to take longer as the quality of the research is likely to be dependent upon the quality of the preparation. It cannot be presumed that young service users will already have the skills to contribute effectively to a research project. It is highly likely that they will need support and preparation for undertaking research, analysing and interpreting the data, writing up the research and disseminating the results. This preparation and training needs to go on alongside the other activities in a young person's life. Lead researchers need to acknowledge that the rhythm of a young person's lifestyle beats at a different rate to that of an adult researcher's. Issues such as friends, partners, education and/or employment will potentially take precedence over the research process. It is also likely that because of these issues the research will have to be undertaken in 'bite size chunks'. It may thus be prudent to over-recruit the number of young service user researchers you may need. This is all likely to result in the research taking much longer and as a consequence costing more and being more resource intensive.

Young service users (as do other service users) require the research training and research materials to be engaging and interesting if they are to become committed to the research project for any length of time. Consequently some stages of the research may be less appealing to young people than others. For example, some young people may find it particularly difficult to engage with the analysis or the writing up which may appear technical, difficult, time consuming and boring. This is potentially the case for those whom schooling is/was not a successful experience. This does not mean that young people cannot participate in the analysis or comment on draft reports but that this might require greater effort and the development of alternative techniques.

One other difficulty that needs to be acknowledged is that this type of research does not suit every researcher. Some researchers, possibly because of personality or life experience, may not be comfortable working with young service users. This should not be unexpected but it is often overlooked. Social workers often choose to specialise in different client groups bounded by age when working in adult teams or child and family teams – it should then not be surprising that social work researchers may do the same.

Employment of Young Service Users

The employment legislation for young people is complex with young people considered to be of compulsory school age until the last Friday in June of the school year in which they reach their sixteenth birthday. Whilst of school age a research sponsor will need to register the research work with the local education welfare department. Forms will need to be completed and consent sought from the parents and the school. It is possible that a school may decide not to give their consent, for example if the young person's attendance is below average. Failure to register may invalidate an employer's insurance policy. There are also time restrictions on how long and when young people are allowed to work in both term time and school holidays. These times also vary with age, although generally children under 14 are not eligible for work. This makes it difficult for young people under the age of 14 to engage in collaborative research although treating them as volunteers may help to circumvent this requirement. Other arrangements in terms of insurance, health and safety and covering expenses would still need to be addressed.

The rules for those no longer subject to compulsory schooling vary between those in the 15–17 age band and those aged 18 and over. A staggered national minimum wage rate applies to this group with the full national minimum wage rate applying to those aged 22 or above. Some organisations pay young people an honorarium for their time and skills (from about £10 to £50 per day in addition to expenses). If under 14 years of age this work is not counted as 'being employed' as the young person is not taxable nor likely to earn enough to start paying tax (Steele, 2003). Those young people in receipt of benefits are expected to declare any income and it is good practice for the research sponsor to remind young people of their responsibility.

Informed Consent

Informed consent was discussed in **Chapter 4** and is usually used to refer to those who are being asked to participate in a research study. In the case of service users it can also be argued that the service users who are being asked to become researchers need to be able to give their informed consent to becoming researchers. The idea of being a researcher can seem very exciting to a service user who may be unaware of the expectations of the role and what it will demand of them. It is thus imperative that young service users who wish to become researchers should be fully informed, as far as it is possible to do so, about what will be expected of them. They are then in a position to opt in, or out, with a realistic view of what is likely to happen. For example, researchers are expected to retain confidentiality and young researchers who are researching their peers need to be very clear that this also includes them and that they cannot go telling their friends what research respondents have said. It is imperative that this informed consent should be checked out at regular intervals to ensure the co-researchers are aware of the expectations in relation to confidentiality and are provided with the opportunity to back out or be supported with any difficult issues. It is also good practice to develop a service user job description that service user researchers can retain to remind them what they have agreed to.

Safeguarding Young Service Users

As France (2004) notes researchers who come into contact with children and young people have a responsibility to protect them from harm, irrespective of whether their contact is professional or informal. This clearly raises issues concerning the limits of confidentiality. For young service users we may wish to debate what the thresholds are and what happens after these thresholds have been crossed. Young service users need to know what the consequences are if they disclose something, how it will be acted on and what that process will be. Ideally this should be dealt with through discussion or the use of potential scenarios but adult researchers need to be aware that the responsibility is theirs to inform the young people. If a young person is unable to sign up to the confidentiality protocol they should be asked to leave the research team or to adopt an alternative role in the process which would not expose them to sensitive material.

For young service users acting as co-researchers the demands of informed consent and confidentiality act both ways. Anything they say, to adult or peer researchers, may be dealt with in this way and anything that is told to them by other young people is subject to the same process.

In seeking to safeguard young co-researchers they should not be asked to interview adults by themselves at unknown addresses or in closed rooms. Interviewing in pairs is advisable not only because of the safety issues but also for providing support and learning. The lead researcher also needs to know where the young co-researchers are and to provide regular supervision and debriefing. McLaughlin (2005) also found that teenage girls needed to interview in pairs when interviewing teenage boys who were more interested in 'chatting' the girls up than answering the questions. As a principled position young co-researchers should be aware that their own safety is paramount and that this is placed above the completion of the research tasks (Save the Children, 2004).

Other practical considerations need to borne in mind including ensuring travelling arrangements are appropriate for young service users, parents or those with parental responsibility are aware of what is happening and that the young service user knows, as much as it is possible to know, about what is expected of them in the research. It is also important that supervision or mentoring has been identified to support the co-researchers throughout the process and that any papers they are asked to read are sufficiently child friendly and free of jargon to make them understandable.

What can be expected of young people will vary in relation to their maturity, age and ability. Researchers need to consider these interrelating issues when deciding what they want young service users to undertake in relation to the research. Young people researching other young people will also need to consider how they can best communicate with other young people. This may require the use of a mosaic approach involving multi-method research styles including the use of computers, pictures, smiley faces and drawings as well as the usual interviewing styles (Langston et al., 2004).

In working with children and young people in the UK researchers need to have a satisfactory enhanced police check. This should also include young people above the age of 16 who are part of the research process. This is especially important if the researchers intend having unsupervised access to young co-researchers or young service users as part of the research. This is not always necessary for other service user groups, but is desirable when working with vulnerable individuals. The police check cannot be a guarantee of acceptable behaviour and research sponsors need to consider how they can minimise the risks and ensure that researchers and young co-researchers behave in acceptable ways.

Conclusion

This chapter has set out to highlight a growing trend in service user involvement in research. Social work can be proud to be at the forefront of this movement. Whilst the chapter has been very much in favour of service users' involvement in research this support has not been uncritical. Service users as researchers should be supported when it would add value or additionality to a research proposal – it is neither always necessary nor appropriate. Before deciding when to involve service users the research sponsor, or lead researcher, need to examine the research question, the types of research methods to be used and the funding in the proposal. If it is deemed that such an approach would benefit from service user involvement the lead researcher then needs to decide at what level(s) and whether this is to include seats on the advisory group or to be involved in the whole research process or parts of it. Having made the decision to involve service users, funding needs to be made available to provide appropriate training, remuneration and to cover research expenses. The lead researcher also needs to be aware that the research is likely to take longer and although the benefits may be potentially greater there is no guarantee that they will be. If you do not have the time to do this properly or the resources to fund such a programme it would be better not to involve service users. Service user research is not a universal solution. Service user research will not solve all the social work researcher's problems but it is another important tool in the social research toolbox and as such it has to be used properly and in the right circumstances to be effective.

Summary

This chapter has sought to examine the issue of service user involvement in social research. The chapter began by identifying the policy and government drivers for the promotion of service user involvement in social work practice and education. This was then developed in relation to social research with the identification of the claims made for the involvement of service users in research and an identification of some of the difficulties related to service user research. Having identified a rationale and value for service user research, issues relating to the philosophical underpinning of this and levels of service user involvement were discussed. In particular a continuum of consultation, collaboration and service user controlled research was critically explored. Overall service user involvement in research tends to be at the lower levels of the continuum whilst service user controlled research is in an emergent phase. Importantly, this section also highlighted that the continuum of service user research carries on beyond consultation to non-participative approaches in the guise of participation. Following the discussion concerning the continuum there was consideration of what the meaningful involvement of service users meant and how a fair return for participation could be ensured to avoid the exploitation of service user researchers.

The chapter then moved onto an examination of the possibilities, difficulties and methodological issues involved in working with young service users as co-researchers. In particular this section focused on issues concerning the employment of young people, informed consent and safeguarding young people.

Suggested Reading

Barnes, C. and Mercer, G. (eds) (1997) *Doing Disability Research.* Leeds: the Disability Press. This collection of articles provides a thought-provoking overview of the development of service user research in 'disability studies', the social model of disability and its impact upon research. This book is currently not in print but can be accessed electronically at www.leeds.ac.uk/disability-studies.

Fraser, S., Lewis, V., Ding,S., Kellett, M. and Robinson, C. (eds) (2004) *Doing Research with Children and Young People.* London: Sage, in association with the Open University Press. A very useful edited book with chapters focusing on some of the key issues in involving children and young people in research.

Hanley, B. Bradburn, J., Barnes, M., Evans, C., Goodare, H., Kelson, M., Kent, A., Oliver, S., Thomas, S. and Wallcraft, J. (2004) *Involving the Public in NHS, Public Health, and Social Care Research: Briefing Notes for Researchers.* Eastleigh: Involve. This is a very useful practical guide giving useful information and detail on involving the public at each of the different stages of the research process. The guide can also be downloaded at www.invo.org.uk.

Kirby, P. (2004) *A Guide to Actively Involving Young People in Research: For Researchers, Research Commissioners and Managers.* Eastleigh: Involve. This is an easy to read guide identifying the benefits and how to involve young service users in research.

7 Research and Anti-oppressive Practice

This chapter will examine what is meant by anti-oppressive practice, why the concept is so important within social work and identify how it impacts upon research and research methodology. Anti-oppressive practice is a contested arena within which students and social workers often struggle to obtain an understanding without either feeling guilty or denying the potential for oppressive and discriminating behaviour.

Dominelli, a key figure on anti-oppressive practice in the UK, has described anti-oppressive practice as:

> A form of social work practice which addresses social divisions and structural inequalities in the work that is done with 'clients' (users) or workers. Anti-oppressive practice aims to provide more appropriate and sensitive services by responding to people's needs regardless of their social status. Anti-oppressive practice embodies a person-centred philosophy, an egalitarian value system concerned with reducing the deleterious effects of structural inequalities upon people's lives; a methodology focusing on both process and outcomes; and a way of structuring relationships between individuals that aims to empower users by reducing the negative effects of hierarchy in their immediate interaction and the work they do together. (Dominelli, 1998: 24)

Dominelli's definition identifies anti-oppressive practice as pervasive impacting on all aspects of social work practice, policy, organisation and delivery. It has an impact at the level of the social worker and the service user, between employer and social worker, between social workers and critically comments on the culture of the agency, their contacts and the social context in which they operate. The role of the anti-oppressive social worker is to highlight, challenge and eradicate social injustice. It can thus be argued that anti-oppressive practice is linked to a longstanding tradition of humanism and social work's long-established alignment with the experience of the underdog.

Discrimination refers to the identification of individuals and groups with identifiable characteristics and behaving less favourably towards this individual or group, than those with favoured characteristics. Anti-discriminatory practice is thus linked to anti-oppressive practice as it challenges workers to behave equitably.

Anti-oppressive practice on the other hand is more pro-active in that it is not only about treating individuals or groups equitably, it is also about conflict and change in relation to the power imbalance between 'superior' and 'inferior' individuals or groups. Anti-oppressive practice, unlike anti-discriminatory practice aims to bring about an equalising shift in the power relations between individuals and groups. Although this distinction between anti-oppressive and anti-discriminatory practice is conceptually distinct in practice the two terms are often used as interchangeable.

One of the key aspects of anti-oppressive and anti-discriminatory practice for UK social workers has been the issue of racism. Anti-racist practice has an important history in connection with the training and development of qualified social workers. In the late 1980s the UK saw the racialised nationalism of Thatcherism defending 'British values' from 'the enemy within' and a promotion of possessive individualism resulting in a xenophobic nationalism (Husband, 1995). Against this background the then social work awarding body, The Central Council for Education and Training in Social Work (CCETSW) official policy statement on anti-racism stated:

> CCETSW believes that racism is endemic in the values, attitudes and structures of British society, including those of social services and social work education. CCETSW recognises that the effects of racism on black people are incompatible with the values of social work and therefore seeks to combat racist practices in all its responsibilities. (CCETSW, 1988 quoted in Husband, 1995: 90)

This policy statement fed into the famous CCETSW Paper 30. Paper 30 codified the Council's system for the approval of the then new Diploma in Social Work (Dip SW) including, as part of the knowledge and skills against which students would be assessed, a student's ability to recognise, understand and confront racism within a multicultural society (CCETSW, 1989, 1991a).

More recently the GSCC Code of Practice (GSCC, 2002) requires social care employers to 'put in place and implement written policies and procedures to deal with dangerous, discriminatory or exploitative behaviour and practice' (GSCC, 2002: no. 4). As part of the Code of Practice employees are expected to engage in:

> 1.5 Promoting equal opportunities for service users and carers and
> 1.6 Respecting diversity and different cultures and values

and to act in such a way as to not:

> 5.5 Discriminate unlawfully or unjustifiably against service users, carers or colleagues
> 5.6 Condone any unlawful or unjustifiable discrimination by service users, carers or colleagues

The National Occupation Standards for Social Work (TOPSS, 2002) also requires that qualifying social workers should be able to 'challenge injustice and lack of access to services' (TOPSS, 2002: 3.b) and 'challenge discriminatory images and practices affecting individuals, families, carers, groups and communities ' (TOPSS, 2002: 6e). The Quality Assurance Agency for Higher Education benchmark statements for social work degrees also highlight this issue of combating discrimination:

> Although social work values have been expressed at different times in a variety of ways, at their core they involve showing respect for persons, honouring the diverse and distinctive organisations and communities that make up contemporary society and combating processes that lead to discrimination, maginalisation and social exclusion. (Alcock and Williams, 2000: 12)

In order to begin to understand the importance of an anti-oppressive social work research stance it is first necessary to consider some of the arenas of anti-oppressive practice. Anti-oppressive practice seeks to help the practising social worker synthesise structural critiques with ethical value bases (Horner, 2003). Anti-oppressive practice focuses on challenging and addressing institutionalised discrimination representing the interests of powerful groups within society. As such anti-oppressive practice is usually considered to cover discrimination on grounds of race, gender, disability, sexuality and ageing. More recently, writers have also talked of the impact of colonialism (D'Cruz and Jones, 2004; Tuhiwai Smith, 1999), the white Irish in Britain (Garrett, 2000, 2003) and for those who live in rural as opposed to urban areas (Pugh, 2003). The discourse of anti-oppressive practice (Dalrymple and Burke, 1995; Dominelli, 1998; Thompson, 1997) endeavours to engage with notions of diversity and difference and we will examine these in relation to race and gender as exemplars of the nature of the debates and insights of anti-oppressive practice.

Macey and Moxon (1996) argue for a move towards anti-oppressive practice away from an anti-racist practice based on practical, theoretical and philosophical limitations of anti-racism but also because of the increasing awareness that many contemporary concerns include the crosscutting nature of issues like gender, class and race. What is required is a move towards an understanding based on the links between various forms and expressions of oppression. This requires examining multiple forms rather than a single dimension of oppression. Issues such as female circumcision cut across black and gender divisions, the rise in religious fundamentalism potentially impacts more on women than on men, both black and white. Anti-oppressive practice recognises the need to maintain a struggle against racism but also raises the possibility of developing alliances across boundaries and learning from others' experiences.

Race and Anti-oppressive Practice

Anti-oppressive approaches include a variety of perspectives that developed in the late 1980s and 1990s in response to a growing concern about ethnic conflict in many Western societies. This included the inner city riots in the UK, which have largely been ascribed to the alienation of young black people. Similar concerns have been raised about the levels of crime and riots involving black people in the USA, over Eastern European refugees in Germany and Italy and a similar unease over refugees from North Africa in France. Many countries face ethnic and cultural conflicts including the position of indigenous peoples, such as the Aborigines in Australia, the Maoris in New Zealand, the Native Americans in the USA and Canada, the Inuit in Canada and gypsies or Romany people in Southern Europe.

Race is 'merely' a social construct that is used as a means of classifying people, to discriminate between people and to exercise control over people on the basis of colour (Everitt et al., 1992). To talk of race as 'merely' a social construct is to signify that race is not real. There is nothing we can point to and say, that is 'race'. There are as many genetic differences between people who share physical attributes as there are between groups divided by race. Race may not exist in any scientifically meaningful sense but this is to neglect that many people act as if it is a fixed objective category.

The conceptual notion of race is intrinsically connected to the idea of racism. Lorde describes this as: 'the belief in the inherent superiority of one race over all others and thereby the right to dominance' (Lorde, 1984: 115). In the United Kingdom racism is about the construction of social relationships on the basis of an assumed inferiority of non-Anglo-Saxon ethnic minority groups. Racism is neither natural nor inevitable and the social construction of race and development of racism are concrete historical processes (Harvey, 1990). This politicisation of human biology into white supremacist terms has enabled those at the top of the hierarchy to construct social relations and impose their power on others. It has also provided a highly visible and simple means to classify people as superior (white) and inferior (non-white). Social relationships organised on white supremacist terms leads to racism becoming internalised and seen as normal. Concentrating on the biological aspects of racism neglects the importance of social relationships whose practices facilitate the oppression of those with visible racial and ethnic characteristics including skin colour, hair type, language and cultural traditions. The politicisation of such characteristics indicate that racism is a social construct whose economic, political and ideological processes represent the means by which the 'superior' group organises hegemony over the inferior group.

Dominelli (1997) identifies a model of racism that includes personal, cultural and institutional elements of such processes and ideologies. Individual racism can be defined as those attitudes and behaviours that indicate a pejorative judgement of racial groups. Unjustified individual racist attitudes lacking institutional support are defined as racial prejudice. Institutional racism consists of the customary routines that ration resources and power by excluding racially 'inferior' groups. In doing so institutional racism pathologies these excluded groups for their lack of success within the system whilst blaming them for their predicament. Thus the interaction of individual behaviour and institutional norms form the dynamic for institutional racism. Cultural racism is centred on values, beliefs and ideas endorsing the superiority of the white culture. As such cultural racism supports and reinforces both individual and institutional racism. In recent years Solomos (2003) notes there has also been a shift from biological superiority to cultural superiority.

Reflexive Questions

At this point you might want to consider why should social work researchers be concerned with race and racism?

What do you see as the implications, if any, for social work if race and racism are not addressed?

Your answer may have included the overrepresentation of black children in the childcare system. Barn (1993) highlighted the significance of race and racism on decisions made by social workers on the care careers of black children. This is a complex issue and as Boushel (2000) comments in discussing the relevance of race for children's social welfare:

> It is now clear, for example, that a specific group of Black children – very young mixed parentage children – may be particularly likely to receive state care, especially if living in white areas (Bebington and Miles, 1989, Charles et al., 1992, Boushel, 1996). In multi racial areas, African Caribbean children looked after by the state seem more likely to return home speedily when fostered by local African-Caribbean families (Barn, 1993). More recently Gibbons et al., (1995) found 'black and asian' children 'overrepresented among referrals for physical injury compared to whites...and underrepresented among referrals for sexual abuse', whilst Farmer and Owen (1995) found particularly high numbers of inconclusive child protection investigations in Black and South Asian families, especially where language problems increased official uncertainty. (Boushel, 2000: 73)

The area of childcare is possibly the best researched in relation to race but the effect of race and racist practices will impact upon every aspect of social work practice, whether this concerns the lack of evidence about the needs of minority ethnic populations (Begum et al., 1994; Butt and Mizra, 1996) or the lack of information available to minority ethnic groups on their entitlement to welfare services (Atkin and Rollings, 1993).

Indigenous Perspectives

One other aspect of the development of anti-racist practice has been the growing emergence of indigenous perspectives. Colonisation by Western peoples following conflicts or wars has resulted in the conquered people's lives, practices and communities becoming transformed into objects of knowledge for the superior colonisers. This has reinforced dominant worldviews as normal whilst marginalising alternatives. From such a perspective discussions about race, class, gender, disability and so on all have their location within the Western world and its models of knowledge production, classification and criteria of evaluation. D'Cruz and Jones (2004: 52) quote Tuhiwai Smith (1999: 80) who claimed coloniser's research practices could best be summed up as: 'they came, they saw, they named, they claimed.'

Immigration, Refugees and Asylum seekers

Following Britain's membership of the European Union, wars in central Africa, the social transformation in Eastern Europe and wars in the Balkans the plight of refugees and asylum seekers has become the subject of a moral panic in the UK. In the 1990s asylum seekers held essentially the same rights as any other citizen. However, this position has increasingly become undermined with the 1993 Asylum Act, 1996 Immigration and Asylum Act and the 1998 Labour White Paper, 'Fairer, faster, firmer' which all sought to minimise the attraction of the UK to 'economic migrants' (which has become a term of abuse), removed access to social benefits and ensured that cash benefits were as small as possible (Ratcliffe, 2004).

Asylum seekers are also dispersed across the country into properties which may be a long way from family, friends or those from their country of origin. As such, asylum seekers are forced into a state of poverty that would not be acceptable for any other citizen in the United Kingdom. It is thus of little surprise that Humphries claims:

> That immigration controls are inherently racist, and no account of 'acting in anti-discriminatory ways' will remedy that basic truth. I also hold that the extreme example of social work's relationship to immigration controls typifies its relationship to social policy generally, in that on the whole it adopts a role of subservience in implementing policies that have exposed the most vulnerable UK populations. (Humphries, 2004: 95)

Although this section has tended to focus on issues of race the reader should be aware that just as 'black' is a problematic term so is 'white' to the point that Garrett (2000, 2003) claims that the racism experienced by the white Irish has been overlooked in the UK and that the mortality rate of Irish males is the only rate that is higher in the UK than it is in the country of origin. Also race is only one aspect of anti-oppressive practice and issues of gender, disability, sexual orientation and class are all central to this approach. Importantly, the reader should remember that people are not uni-dimensional; simultaneously we are also male or female; black or white; disabled or non-disabled; heterosexual, gay/lesbian/ bi-sexual or trans-gendered; upper, middle or working class and so on. We all have more than one defining identity characteristic. Some of these attributes will imply privilege and power whilst others may render us oppressed. What about the 'structural position of a white, working class, homeless male with that of a black barrister' (Macey and Moxon, 1996: 303)? As McDonald and Coleman (1999) suggest hierarchies tend to be unworkable in practice and need to be understood as component parts of an oppressive system.

Associated with this criticism is the lack of attention given to the social division with which social work is most closely associated – that of becoming service users and the receipt of social work or social care support. In particular Wilson and Beresford (2000) highlight the failure of social work to address the oppression minority group members experience as service users. In doing so they highlight five concerns about the current operation of anti-oppressive practice as it relates to service users. In particular they highlight:

- Appropriating users' knowledges;
- Using and reinforcing inherently oppressive knowledge about service users, for example, a bio-medical model of madness and distress;
- Reinforcing constructions of 'clients' as passive and low status by controlling ideas of 'anti-oppressive practice' and knowledge of 'oppression';
- Mirroring and masking traditional professional power;
- Providing continued legitimation for controlling and problematic social work practice. (Wilson and Beresford, 2000: 569–70)

For Wilson and Beresford (2000), social work, far from being part of the answer is part of the problem.

Healy (2005) also refers to conflictual nature of anti-oppressive practice citing Thompson's (1997: 159) assertion that the prefix 'anti' is very significant as it symbolises the fight against powerful forces. She is concerned that such a view polarises postures into 'them' and 'us' positions, neglecting the reality of social work which is often practised in the margins and grey areas of society where negotiation and compromise are essential. To arrive at a situation with preconceived notions will hamper and restrict our ability to listen and to work creatively with a range of stakeholders on agreed solutions.

Anti-oppressive Research

> ### Reflexive Questions
>
> At this stage of the chapter you should reflect upon your position in relation to oppression and write down some of the key messages you have learnt. Having done this you now need to consider how these messages help to inform the need for an anti-oppressive research approach.

In reflecting upon your position you may have identified some of the things below:

- Oppression exists.
- Am I part of the problem or the solution?
- Anti-oppressive practice is contentious.
- Anti-oppressive practice makes me feel uncomfortable.
- Anti-oppressive practice exists at a number of different levels.
- Anti-oppressive practice is embedded in social work values, but is more difficult to see in practice.
- Anti-oppressive practice confronts me with issues of power and discrimination.
- Potentially, I am both an oppressor and oppressed.
- Anti-oppressive practice is about challenge – to oneself, one's agency and society generally.
- Concepts like 'race' are mere social constructs that have major life influencing implications for us all.
- Anti-oppressive practice is about empowerment.

Your list may be much longer than the list identified above although, hopefully, it has included some of the comments above. As already noted anti-oppressive practice, as is research, is driven by moral and ethical value codes which explicitly challenge sources of privilege or power that are based on unjustifiable differences. Butler's (2002) code of research ethics adopted by the British Association of Social Workers states:

> Social work and social care researchers must not tolerate any form of discrimination based on age, gender, class, ethnicity, national origin, religion, sexual orientation, disability, health, marital, domestic or parental status and must seek to ensure that their work excludes any unacknowledged bias. (Butler, 2002: 245)

Within the code there are references to 'working together' with 'disempowered groups' in ways, which 'aim to promote social justice' (Butler, 2002: 245). Social work research ethics can thus be seen as promoting a distinctive attitude towards social research that seeks to stake out an identifiable professional position. As identified in **Chapter 4**, neither social work nor social work research are ethically neutral and both aim to empower whilst promoting the welfare of service users and ensuring access to social and economic capital on equal terms with other citizens. Some communities are still under-represented in research, for example, Chinese communities and newly arrived populations. No single approach will meet the needs of all groups. The circumstances of Chinese older people may differ from that of Afro-Caribbean older people or Anglo-Saxon older people. Also as with older people generally, each person's circumstances will be unique. It is also possible that within such categorisations other forms of discrimination, like being disabled, may be marginalised or under-represented in research. It should also be remembered that whether black and ethnic minority people are born in the United Kingdom, or arrived from elsewhere, may significantly impact upon their personal experiences which may not be reflected in research (Hanley, 2005). In contrast, some communities are over researched – Butt and O'Neill (2005) found that this was the case in a study of black and ethnic minority older people. They claimed that researchers had often asked the same questions and produced the same findings that were evident 15 years ago. In particular, they did not want research for its own sake, but wanted action that would bring about change.

In adopting an anti-oppressive research approach it could be argued that the researcher is abandoning the notion of objectivity. As Barn (1994) comments:

> Are we abandoning the notion of objectivity if we espouse the principles of anti-discriminatory research? Are we allowing our prejudices, our biases, our pre-conceived notions to come in the way of 'proper academic research'? If we begin to answer these questions, we need to address the whole notion of objective research. Furthermore, if we accept that there is a dialectical relationship between theory and ethnography, objectivity as a concept of purity begins to hold little meaning. (Barn, 1994: 37)

This then raises another issue as to how we are to assess research and in particular research concerning anti-discriminatory research. Hammersley (1995) has highlighted the controversial research of Foster (1990) who studied a multiethnic, inner-city school which was formally committed to anti-racist education and found there was little evidence of racism amongst teachers or of school practices that indirectly disadvantaged black students. This was irrespective of whether you examined how the students were treated in the classroom or how they were allocated to ability groups. These conclusions challenged the perceived wisdom and resulted in a predominantly negative response. Foster (1991, 1992) provided a robust response to his critics concerning the challenges over his methodological and epistemological assumptions. In particular Foster's research challenged his critics' assumptive worldviews which took for granted that racism was endemic and institutionalised within society and therefore Foster could not have found a situation where this was not so. In other words he challenged a core assumption of the 'anti-racists' position who were then placed in a incommensurate position whereby by definition Foster's research must have been flawed as its results did not fit into their paradigmatic worldview (see **Chapter 3** for a fuller discussion on Kuhn and paradigms). To accept Foster's research was not flawed would mean their core assumptions on racism and its ubiquitousness would need to be reconsidered. In order to deal with such a situation Hammersley (1995) suggests that where disagreements arise there needs to be an attempt to resolve the situation based on judgements about the plausibility and credibility of the evidence. This process would then be able to identify the acceptance of one of the two original positions, or more likely the development of a third. In order to promote the opportunities for an agreement to be reached Hammersley (1995: 76) proposes the adoption of the following five norms:

- The overriding concern of researchers is the truth of claims, not their political implications or practical consequences.
- Arguments are not judged on the basis of the personal and/or social characteristics of the person advancing them, but solely in terms of their plausibility and credibility.
- Researchers are willing to change their views if arguments from the common ground suggest that those views are false.
- Where agreement does not result, all parties must recognise that there remains some reasonable doubt about the validity of their own positions.
- The research community is open to participation by anyone able and willing to operate on the basis of the first four rules; though their contributions will be judged wanting if they lack sufficient knowledge of the field and/or of relevant methodology. In particular there must be no restriction of participation on the grounds of religious or political attitudes.

Hammersley has tried to identify a structure based on the assumption that researchers are reasonable people committed to the development of knowledge. Topics like anti-discriminatory research put these assumptions to the test challenging both the researcher and the practitioner who uses the generated knowledge in their practice.

Feminist Research Practice

Given the position of women in social work as practitioners, students and service users it is also worthwhile examining some of the issues involved in feminist approaches to social work and social work research.

Reflexive Questions

The majority of qualified workers in social work in the UK are women. The majority of those undertaking qualifying training in the UK are women. Why do you think this might be so?

What, if any, do you think are the implications of this for social work practice and research?

Feminist perspectives have played an important role in the development of social work practice. However, it should not be assumed that there is one feminist position or approach. Harvey writes that:

> There are a number of different feminist views about the nature and mechanisms for the oppression of women. A lot of prefixes have been added in various combinations to feminism: socialist, Marxist, bourgeois, radical, positivist, idealist. Unfortunately, these labels have not all been used to mean the same thing, nor are they mutually exclusive. More profoundly, the theoretical positions embodied in different perspectives are not necessarily distinct. (Harvey, 1990: 106)

It is thus quite difficult to tie down the feminist perspective and instead we identify feminist approaches. Payne (2005: 253) identifies five different feminist perspectives:

- Liberal feminism – seeks equality between men and women, particularly in the workplace, caring and family responsibilities. For the liberal feminist the answer to inequality lies in promoting equal opportunities by legislation, challenging and changing social conventions and modifying the socialisation process so that children do not accept gender inequalities.
- Radical feminism – highlights patriarchy, roughly translated as male dominance. This view values and celebrates the differences between men and women and seeks to promote separate women's structures within existing organisations.

- Socialist or Marxist feminism – women's oppression is viewed as part of structural inequality within a class based social system. Women's role is conceptualised as reproducing the workforce for capitalism, childcare and domestic tasks. Oppressive relations need to be studied alongside other forms of oppression and class based structures so that diverse interests can be met appropriately.
- Black feminism – this came about as a reaction to white feminists who saw the omni-relevance of gender for all women. This view was challenged by black women who experienced racism as their primary defining characteristic. Black feminists also claimed a heightened awareness of oppression as opposed to white feminists due to their experience of racism. Black feminists look towards connections with other socially oppressed groups to campaign for social justice.
- Postmodern feminism – this view identifies the complexities of social relations identifying how discourses help to shape social assumptions about the role and expectations of women. In particular this view adopts the postmodern concern to question categories and to examine social relations in terms of surveillance and discipline in relation to families and social workers.

Each of the above approaches is open to challenge and criticism. Liberal feminism can be criticised for ignoring the differences in interests between men and women and promoting equal opportunities. Equal opportunities may not be realised if other aspects of inequality prevent women from grasping the opportunities provided. Radical feminism with its focus on gender differences and the common experience of women is criticised for ignoring the diversity between different groups of women. The black feminist position can be criticised for its view that black women have a more complete view of social reality precisely because of their disadvantaged position. Such a view neglects the position of women who may be disabled or lesbian who may claim a similar primacy of position leading to a potential hierarchy of oppressions. Socialist feminism is open to criticism by radical feminism for its failure to explain patriarchy in terms of class and economic oppression. Postmodern feminism can be criticised as it sees gender as a social construction as opposed to a sexual identity.

Each of the differing approaches to feminist scholarship can be seen to contribute to an increased understanding of gender or gender relations. Patriarchy is a key issue for understanding feminist theorising. Roughly translated it means male dominance. Within feminist approaches patriarchy is fully assumed, identified as negative and in need of being changed. This also includes social research methods that are sometimes described as being 'malestream' and reflecting a 'masculine' view of the world. Feminists on the other hand seek to emphasise personal experience in that the 'scientific method' and in particular positivism neglects the researcher's embeddedness in the socio-political world and neglects the importance given to gender relations in structuring this knowledge. At root, the 'male paradigm' denies the relevance of the personal. The personal is not a constituent of knowledge. Feminists reject this view and affirm the validity of direct experience. For feminists knowledge comes from rediscovering links between the personal and political, affirming women's ways and experiences of knowing and building collective insights through consciousness-raising, by forming groups to share experiences and providing mutual support.

Feminist social research approaches use a specific sub-set of methods and/or select a particular set of topics with the aim of challenging methodologies developed by men and improving the position of women (Payne and Payne, 2004). In the past research has studied men, not women. Most of this has tended to focus on the public sphere, for example work or crime; the private sphere was overlooked or not considered worthy of research time. This meant that women became invisible as they were often unrepresented and restricted to the private sphere of family and children. Worse, often social scientists wrote as if what applied to men was not only masculine but also 'universal', the use of 'people' often meant 'men'. This viewpoint also promoted the notion that social regularities were gender blind – the use of 'people' was often based solely on samples of men. Feminists highlighted these shortcomings and sought to understand the daily lives and experiences of women. In particular feminists argued that this required a very different approach to research and this helped promote the development of approaches like ethnomethodology, autobiography and narratives. These approaches promoted a view which saw the need to relate to other women as 'subjects' not 'objects' moving towards a more collaborative, non-hierarchical and inclusive practice (Oakley, 1981). In seeking to capture women's experience feminist research argued that feminist accounts should also include 'feelings, other interests and even unconscious beliefs' (Payne and Payne, 2004: 91). To do this feminist research approaches knowledge generation from 'below' as opposed to from 'above'. In the search to be non-hierarchical research seeks to serve the interests of the exploited and oppressed groups, especially women and it would be a contradiction to do this with a traditional social research perspective or view from 'above' which would be seen as directly oppressive and inhibiting women's liberation.

Feminist research has rightly raised the issues of women's lives and the private sphere to the level of worthwhile research. It has helped us to understand issues, such as teenage pregnancy that has traditionally been seen as stopping teenage girls under 18 becoming pregnant with little regard to the male's role in the process. Feminist research reminds social workers of the position of many of the women whom they seek to work with either as women, parents or carers. It also offers lessons for male social workers and social researchers in how to address and work with women's perspectives and the importance of gender for structuring social knowledge. In particular it offers a timely challenge to patriarchy and a more holistic view of social issues. However, in giving greater priority to women's issues it is in danger of mirroring the mistakes of previous research by neglecting the role and needs of men. It is very helpful to understand the experience of women who suffer domestic violence. However, we do these women an injustice if we do not also seek to understand why some men behave in this way and also seek to identify interventions that can both protect victims and address violent behaviour.

Given the gender bias in social work it is not surprising that feminist approaches have appeared attractive both in practice and in research. However, Orme (2003) sounds a note of caution – women suffering oppression cannot always be given priority, or even believed. In issues of safeguarding children the child's needs are paramount even if there is a conflict between their needs and their mother's. This is irrespective of whether this will result in the further oppression of the mother. In research terms feminist approaches have helped develop qualitative methodologies emphasising dialogue, mutuality and egalitarian relationships between researchers and those being researched. Feminist research and research practices have helped the development of anti-oppressive research and have helped to ensure that the daily life of women is not forgotten. In particular feminist research has helped to problematise gender and gender relations ensuring these issues are legitimate areas of concern. Central to both the discussions on race and gender has been the importance of empowerment which we turn to next.

Empowerment

As mentioned above anti-oppressive research (and practice) seeks to build partnerships with service users and oppressed communities to promote empowerment, but what is empowerment? Empowerment is a contentious term that has become fashionable within health and social care. It has though been open to criticism over its contradictions (Humphries, 1996) and has been described as an abused and devalued term by Croft and Beresford (1989).

> [Empowerment] has been taken to mean challenging service user's exclusion and disempowerment. Two aspects to empowerment are regularly identified: personal and political empowerment. Personal empowerment is concerned with strengthening the individual's position, through capacity and confidence building skills, and assertiveness training, to be able to gain more power. Political empowerment is concerned with seeking to make broader change that will increase objective power (political, cultural, social and economic) available to people. A particular appeal of the idea of empowerment is the way that it can unite these two concerns: individual and social transformation. (Beresford in Hanley, 2005: 15)

It is also worth noting that service users claim that people can only empower themselves, neither researchers nor social workers are in a position to do this for them.

Anti-oppressive Research Practice

The Research Sponsor

A researcher concerned with researching in an anti-oppressive way will pay attention to who funds the research. Just as a public health researcher is unlikely to accept a commission from a tobacco company it is also likely that those researchers adopting an anti-oppressive research practice will only wish to secure funding from organisations that wish to address anti-oppressive issues or, at least are open to this possibility. This may restrict funding sources but is also likely to avoid undue future conflict.

The Research Question

In adopting an anti-oppressive research approach the first challenge for the researcher is to consider their research question. Is it a question that seeks to empower or one that will further discriminate or oppress particular groups? Researchers in the future might wish to consider whether they should begin with what black and ethnic minority communities may see as a priority. So, instead of focusing on what people's experiences are of service provision they may wish to begin with what the research subjects consider to be important. If we do not do this we run the risk of only identifying those priorities that are important to the researcher and research sponsor and run the risk of further oppressing research subjects. In a similar vein it is not enough to search for clues as to why black people are more likely to be diagnosed with schizophrenia without asking what are the social processes which contribute to a disproportionate number of black people being diagnosed as schizophrenic. Or, in focusing upon domestic violence to stop blaming the victim and begin asking, not, why do battered women stay, but asking what are the factors and conditions that make battering possible or even acceptable. A common feature of anti-oppressive research practice is to challenge common conceptions, for example children in residential care are all too often focused upon as 'a problem' without any acceptance or consideration of their strengths or the social relations that led to them becoming looked after.

Practice Example

Barn (1994) outlined the methodological conflicts and constraints she experienced when she attempted to undertake an anti-discriminatory research study into the processes involved in the admission of children into care in two local authorities. Barn, a black researcher, attempted to consolidate issues of class and socio-economic disadvantage alongside issues of race, ethnicity, culture and religion into the research framework. From the outset she experienced problems with the research where after nine months of negotiation she had failed to gain access to one of the local authorities who felt her research design was too ambitious, that social workers felt their practice was being put under scrutiny and who queried the motives of higher management for agreeing to such a study. Social workers also felt that it would be inappropriate to interview service users and if they were to be interviewed the social workers should retain a veto as to who was seen. This raised issues as to when modifications and adaptations in real world research become manipulation by powerful interest groups. In the end this authority withdrew from the study. Another two authorities were approached, but they also refused access. The research thus progressed in only one authority, Wenfold.

In interviews Barn faced the problem of whether respondents were telling the truth or not. She also considered the issue of whether she could obtain an accurate picture of white social workers' perceptions especially as they might fear expressing politically incorrect sentiments. In response to black and white parents she felt the issue was less bound by race as they did not have to fear giving an 'inappropriate answer'. Age, and the question of whether she had children, seemed to be more important than race and gender. The social workers also controlled access to the children and this created difficulties as the social workers felt it was their job to protect the children. She also felt that social workers would be obstructive in cases where they did not have a good relationship with the child to ensure the service user did not have an opportunity to express their view. Barn also comments that not everyone saw her as an outsider and that she was often mistaken as an employee and needed to assert her autonomy as a researcher.

From Barn's research it is clear how powerful agencies can shape and transform the nature of the research design to the point of denying access. In trying to act in an anti-discriminatory way a researcher not only needs to address the issues that every researcher faces but also maintain their commitment to an anti-discriminatory research practice and tread 'the path with much caution, diplomacy and tact' (Barn, 1994).

Barn's example demonstrates some of the difficulties in engaging with anti-discriminatory research. Those with power are unlikely to wish their power to be threatened or questioned and as such will operate in a way that modifies the research to limit its impact or undermine its operation. Barn's research also reminds us that although as social workers we are all likely to agree that social work should address inequalities and promote social justice it becomes a more difficult question when it is us and our practice that is under scrutiny – however laudable the aims of the research.

In considering anti-oppressive research the researcher needs to consider a number of practical issues including ensuring any venues are culturally appropriate and easily reachable. Interpretation and translation services may be required if the research respondent's first language is not English. It is important to discuss the research with the interpreter before it takes place and to avoid using family members for this activity. Whilst it may be convenient to use a family member the nature of the research may be sensitive to the respondent and they may not want their business known to other family members. Also, a family interpreter may reinterpret the respondent's responses as to what they believe to be acceptable. This need for interpreters can in turn increase the costs of research. This position is further exacerbated as Hanley (2005) has noted funding to involve black and ethnic minority communities in research is often under-resourced.

Anti-oppressive Research Methodology

Anti-oppressive research is more than the need to avoid sexist, ageist, disablist, racist, homophobic and other types of discriminatory language in the use of questionnaires, interviews and writing of reports. Anti-oppressive research also implies the need to treat the research subjects as people and not as objects. This may require a collaborative, participatory or emancipatory research approach. Everitt et al. (1992) identify three reasons as to why collaborative research may be beneficial to both the researched and the researcher alike. By giving credibility to the views of those less powerful the researcher will gain access to the experiences of being discriminated or oppressed and of the behaviours and perspectives of those who oppress. Secondly, by having due regard for people in the research process research subjects will become less suspicious of research and engage more honestly. It is also possible that by having one's views and experiences validated by research will be experienced as empowering, thus increasing the research subject's willingness to share. Thirdly, those with the experiences the researcher wishes to explore are more likely to be aware of what questions matter and how answers should be interpreted.

Lincoln and Denzin (1998) identifying the crisis of representation of the research subject argue:

> Who is the Other? Can we ever hope to speak authentically of the experience of the Other, or an Other? And if not, how do we build a social science that includes the

Other? The short answer to these questions is that we move to including the Other in the larger research processes that we have developed. (Lincoln and Denzin, 1998: 411)

In anti-oppressive research this will imply not only working in ways that knowingly do not harm the research respondent, but also seek to engage them effectively in the process. Heron (1996) identifies this requirement as the demand for a participative paradigm of social research including both epistemic and political concerns. The epistemic suggests an epistemology that asserts the participative relation between the knower and the known, or Other, and where the other is also a knower, between knower and knower. Knower and the known or the Other are not separate in this relationship, with the degree of participation being partial and open to change. Methodologically such an approach commends the validation of research outcomes through the 'congruence of practical, conceptual, imaginal and empathic forms of knowing among co-operative knowers, and the cultivation of skills that deepen these forms' (Heron, 1996: 11). Politically this view values human development as intrinsically worthwhile and achievable through participative decision-making.

Clifford (1994: 104) summarises what he sees as the key features of an anti-oppressive research. An anti-oppressive research strategy would be:

- Anti-reductionist and historically specific – placing any explanation within a historically specific context and avoiding explanations based on biology, psychology and economics.
- Materialist – relating any explanation to the divisions of material wealth and power in society.
- Combining the personal and political.
- Thoroughly analysing 'difference' – placing individual and groups within all the social divisions.
- Internationalist – being aware of the wider contexts which affects us all either indirectly or directly.
- Reflexive – the researcher or observer is accountable for the methods used and the knowledge claimed which does not exist outside the framework at some value-free point, but is part of the social action and thus part of the research.

Research Dissemination

An anti-oppressive research approach would also consider the nature of the dissemination process. This would take into account how the research would be disseminated to those who were party to its production. Issues like the use of plain English, trade journals, translated versions into the subject's first language, videos, DVDs and presentations all need to be assessed to build an effective and accessible dissemination strategy. Ideally this strategy should be identified at the start of the process, appropriately funded and checked out with the research subjects as the research progresses. On top of this the researcher needs also to ensure, as far is possible, that what they publish will not result in future or present harm to the research subjects.

Summary

This chapter has sought to identify the importance of anti-oppressive practice to social work particularly using the exemplar of race and showing how this can be translated into research. The chapter began noting that anti-oppressive practice is a historically contentious arena and although usually associated with class, race, gender, disability and sexuality is not limited to these defining characteristics and may include ageism, colonialism, urbanism and other defining characteristics by which one group becomes oppressed by another. The challenge here is not to develop a longer list than anyone else but to be able to see how such social constructs act as defining characteristics, which are often left unseen and unchallenged, and how these impact upon services and (potential) service users.

Anti-oppressive practice and research needs to be considered critically. As this chapter has shown individuals are not uni-dimensional, people may be white, female, disabled and so on, some of whose cross-cutting characteristics will interact with each other. Concepts like 'blackness' and 'whiteness' have been shown to be problematical and potentially restrictive. We should thus avoid seeking to establish a hierarchy of oppression or a mathematical additionality of oppressions as some of these characteristics will impact differently in different circumstances.

Anti-oppressive research implies challenge at the personal, cultural and institutional levels. Anti-oppressive research is not only about the research question and outcomes, but importantly, how the research is undertaken. To undertake an anti-oppressive research project in a non-anti-oppressive way is a contradiction in terms. This is not to say that traditional research projects cannot be used to help develop and inform anti-oppressive research processes and insights. It is to say that it is not possible to espouse an anti-oppressive research affiliation whilst behaving in an oppressive way. Anti-oppressive practice is a key concept for both social work practice and research; it covers contested terrain and is associated with challenge and rebalancing social relations.

Suggested Reading

Dalrymple, J. and Burke, B. (1995) *Anti-Oppressive Practice: Social Care and the Law.* Buckingham: Open University Press. Although this book focuses largely on using the law, it has become a classic highlighting the principles of anti-oppressive practice and empowerment of service users throughout.

Healy, K. (2005) *Social Work Theories in Context: Creating Frameworks for Practice.* Basingstoke: Palgrave Macmillan. A thought-provoking book which includes a considered critique of anti-oppressive practice.

Humphries, B. and Truman, C. (eds) (1994) *Re-thinking Social Research: Anti-Discriminatory Approaches to Research Methodology.* Aldershot: Avesbury. A collection of articles addressing key issues involved in the why and how of anti-discriminatory research processes.

Truman, C.E., Mertens, D.M. and Humphries, B. (eds) (2000) *Research and Inequality.* London: UCL Press. A collection of papers focusing on research and the challenges involved in researching oppressed groups in a non-oppressive way.

Wilson, A. and Beresford, P. (2000) 'Anti-oppressive practice: emancipation or appropriation?', *British Journal of Social Work*, 30(5): 553–74. A challenging critique of anti-oppressive practice in operation and how service user groups experience it.

8 Interdisciplinary Contributions to Social Work and Social Work Research – the New Orthodoxy

> Social work services don't have all of the answers. They need to work closely with other universal providers in all the sectors to find new ways to design and deliver services across the public sector. (Roe, 2006: 8)

The above quote is one of the three key messages highlighted by the Roe commission into the future of social work in Scotland and this chapter sets out to highlight the importance and growing contribution of interdisciplinary practice and research. There is a growing awareness of the interdependence of services and this has been a feature of service development worldwide including Australia (Hugman, 2003; Rogers, 2004), America (Pew Health Professions Commission, 1995; Stone et al., 2004) and the UK (Davey et al., 2005; Leathard, 2003; Trevillion and Bedford, 2003). In the UK, since 1997, interorganisational working has moved quickly up the policy agenda where it has been supported by a raft of legislation, special initiatives and funding. This has ranged from dealing with complex and 'wicked' social problems such as crime and social exclusion, to the experiences of particular service user groups including people with mental health needs, people with learning difficulties, older people or people with physical disabilities (Glasby and Lester, 2004). The emphasis on joined-up-thinking for joined-up-problems can be seen in initiatives like the Crime and Disorder Act (1998) that led to the creation of the youth crime service and the establishment of interorganisational, multi-agency staffed Youth Offending Teams.

Interdisciplinary working in these contexts refers to:

> situations where workers belonging to one professional discipline practise beyond their own boundaries in partnership with professionals from other disciplines (Ovretveit, 1993, 1997). This kind of practice is sometimes known as multi-disciplinary practice or inter-professional practice (Leathard, 1994; Higham, 2001:21)

Of particular importance to us from Higham's definition are the notions of a profession and the notion of a discipline.

Reflexive Questions

Before going to examine the notion of a profession, can you consider what you believe a profession to be?

Can you name three professions who meet your definition. Would social work be included within your definition?

Why is this so?

What about the soldiers, police officers, nurses, publicans, doctors, lawyers and footballers?

Is Social Work a Profession?

In today's society many groups claim to be 'professional', for example, the Army invites you to come and 'join the professionals'. We talk of professional footballers or the Police whilst many people still look towards the time-honoured professions: doctors, lawyers and clerics.

Traditionally, Hugman (1991) notes that the notion of a profession has been developed from a delineation of 'traits' or characteristics that were held to constitute a profession. Through claiming professional status the medical profession has been able to successfully organise itself to maintain a high standing in society and protect a high level of material reward. This has been based on three distinct characteristics. It has restricted entry to the profession via a strict set of criteria or qualifications. It has a professional association who monitors and controls the conduct and performance of its members and it is generally accepted only those who it licenses are qualified to practise medicine (Giddens, 2001). In so doing the profession is able to exclude unwanted individuals whilst at the same time enhancing the reputation and market share of their own members. In contrast occupational groups like social workers, nurses or physiotherapists have been viewed as semi-professions or bureau-professions (Payne, 2000). These occupational groups are seen as aspirating professions but lacking the autonomy of practice.

The trait approach has come under criticism as, first, it ignores that the claims for professions like medicine or the law should not just be accepted but should be opened up for questioning rather than being seen as benchmarks for others to achieve. Secondly, the extent to which a trait is assigned to a profession is in reality a claim to exclusion and exclusivity, which needs to be challenged as opposed to being accepted as a social fact. Thirdly, such an approach ignores issues of power and how certain occupational groups can muster sufficient weight and support to call themselves, and be called by others, professionals. The issue of whether an occupational group is a profession or not is a debate about power whereby occupational groups benefit from asserting themselves as professions in order to exert self-control.

It could be argued that social work is now a profession in that entry to the profession is controlled through the social work degree qualification. The care councils have taken over the role of controlling and registering those who may and may not be admitted to the profession. Also, it is only those who have the appropriate qualification and are registered who can legally call themselves social workers following the protection of the title of social worker in April 2005. On such a tri-partite definition as suggested by Giddens (2001) social workers could justifiably claim to be members of a profession.

Others may point out that in 'being a profession' traditionally has indicated a model of individual practitioner responsibility. Doctors are deemed to hold 'clinical responsibility' for patients and lawyers are deemed to be 'individually responsible' for their own practice. Clearly social work has not reached these degrees of individual responsibility being located primarily within statutory or voluntary bodies. It is though worth noting that the levels of individual responsibility within the traditional professions have also been reduced at the same time as social work's has been increased. The medical governing bodies have come under increasing scrutiny with the British Medical Association being criticised by Dame Janet Smith's public inquiry into the deaths of Dr Shipman's patients (Smith, 2004). Increasingly, lawyers are required to specialise and are employed in non-legal organisations. In all these cases, consumer organisations, the development of complaints systems and increasing social pressures has reduced the autonomy of the traditional professions.

Our understanding of the nature of professions needs to become more complex. We need to consider autonomy and responsibility more broadly along with concepts of power and the ability of an occupational group to persuade other occupational groups to respond to them as a profession.

Is Social Work a Discipline?

It is rarely asked what it means to be a discipline or whether social work is a discipline or not. Deetz (1993) in writing about communication studies has argued that there are at least three different approaches to the notion of an academic discipline. The first of these (Discipline 1) suggests that a discipline represents a world organised into academic units or departments. On this criterion social work is obviously a discipline as it is represented in many universities throughout the world as a separate department or unit.

Discipline 2 reflects a world organised around topic interests. Again social work can be defined as a discipline as it is seen to be a distinct phenomenon with clear topic interests. Discipline 3 represents a world organised around competing modes of explanation. This is where social work struggles – it could be argued that there are just too many competing forms of explanation for what is taken to be the same phenomena. In this sense the future of social work as a full discipline commanding of respect and productive in research terms relies heavily upon the failures of sociology, psychology, economics and so on to deal effectively with explaining social phenomena whilst also providing direction for social workers and social care practitioners. To the extent that these disciplines fail to account for the nature of social work there is then a danger that social work becomes translated into more than just underpinning social care, but becomes reconceptualised into something that every human being undertakes in negotiating reality. This requires social work to address questions of ontology and epistemology to establish its intellectual core and avoid being at the whim of other academic disciplines. Social work thus becomes an activity that is located in the home, the street and in public and private life. Although its activities are social, psychological and economic it cannot be reduced to any one of these disciplines.

Importantly, it should be noted that the case for social work research is not limited to the activities of social workers, but represents the underpinning of social care – social work skills, knowledge and values are relevant to the wider social care audience and 'we must not confuse "social work research", with "research on what social workers do"' (Marsh and Fisher, 2005).

The Discipline of Interdisciplinarity

It may seem strange to be discussing whether social work is a discipline when as we have already noted there has been a growth in interdisciplinarity accompanied by an expansion in multiprofessional teams and a general acceptance that this represents the way forward for services in the future. In reply, it is probably more important than ever that social work is able to articulate its uniqueness, values and core activities. To engage in multiprofessional teams or interdisciplinary ventures without being clear about the unique contribution of social work is to run the risk of being colonised by other more articulate or older professions. There is thus a paradox, in seeking to work or research in multiprofessional teams, a strong professional identity is required to maximise the contribution of the individual disciplines or to risk the team being reduced to those who shout loudest. Conversely too strong a professional identity is likely to lead to disruptive arguments concerning territory and boundaries.

Alongside this growth in interdisciplinarity Delanty and Strydom (2003) have noted a decline in disciplinarity. The natural, human and social sciences can no longer be so sharply separated from each other. The natural and social sciences share many similar areas of concern including the 'environment', the 'body' and 'risk'. The rise of these new discourses has demanded a new complexity and interrelationships. Also, there has been an increasing trend within social sciences to become more and more interdisciplinary. Research has become increasingly driven by problems (e.g. the reduction of crime), ethical political concerns (e.g. asylum seekers) and policy-directed programmes (eradicating poverty) than by single disciplinary specific traditions. It would be premature to say that there is a crisis in disciplinarity but it would also be naïve to ignore the current trend and unquestioned acceptance of interdisciplinarity.

Interdisciplinary Practice

In seeking to address individual, family, community and society's problems social workers come into contact with a wide range of professions. These professions potentially include health visitors, teachers, probation officers, psychologists, education welfare officers, GPs, nurses and paediatricians in relation to child protection; community psychiatric nurses, GPs, gerontologists, psychiatrists, nurses, occupational therapists and housing officers in relation to elder abuse. These groups may come together in a number of ways. Traditionally they have been based separately and have come together to focus on an individual problem or concern. More recently, following Section 31 of the 1999 Health Act NHS bodies and local authorities have been given the 'flexibility' to respond to user-focused services by joining up services or by developing new co-ordinated services as well as working with other organisations. These have led to the formation of learning disability trusts, mental health trusts and more recently to children's trusts. This has led Higham (2001: 21) to comment that social workers in the future will either practise in interdisciplinary teams or work in partnership with professionals from other occupations, much more so, than with colleagues from their own profession.

Reflexive Questions

Before we go on to identify the benefits of interprofessional practice you should try to identify three things that you think would be improved through interprofessional practice. What would you see as the potential problems in achieving these benefits?

Would you want to work in an interprofessional team?

If so why?

The benefits claimed for interdisciplinary practice teams include (Hudson, 2002; Leathard, 2003; Lymbery, 1998; Torkington et al., 2003):

- Reduces fragmentation and duplication in an increasingly complex policy and welfare context.
- Reflects the government policy agenda on 'joined-upness'.
- More effective use of staff.
- More effective use of resources.
- Effective service provision.
- A more satisfying work environment in terms of service provision and staff support.
- Recognition that professionals can have as much in common with members of another profession as they do with their own.
- Effective interprofessional working can help to meet the collective goals of individual organisations while providing a better overall quality of service.

All of these positives are very persuasive, but Hudson (2002) has claimed that a closer reading of the literature on interprofessionality has tended towards a more pessimistic outcome. Hugman (2003) in reflecting on the experience in Australia has commented that with all the attempts to promote interprofessional developments in the Australian context boundaries between professions have reformed rather than been reduced:

> What are the incentives for any of the health and social welfare professions to move beyond their 'tribal circles?' when the very understanding of the benefit to service users is seen as differently from within these separate perspectives (and attempts to

> weaken boundaries have been experienced as a collusion with the attack on professionalism per se from economic rationalism). Until that question is addressed we may expect that the 'tribal' circles to remain strong and hence to risk continuing to 'go round in circles'. (Hugman, 2003: 65)

It can thus be argued that although interprofessionalism promises many benefits these benefits have been difficult to realise. One area that has been identified as a potential driver for change is well-organised and structured programmes of education and training to support interprofessional practice. Barr (2003) claims that professions work better together when they learn together.

In a review of health and social care in the UK it was claimed that interprofessional education will:

- Enhance motivation to collaborate between professions.
- Change attitudes by contact and countering negative stereotypes.
- Cultivate interpersonal, group and organisational relations.
- Establish common values and knowledge bases.
- Reinforcing competence by defining outcomes in terms of competencies required for collaboration (Barr et al., 1999).

Zwarenstein et al. (1999), using the guidelines for systematic reviews developed by the Cochrane Collaboration, concluded that no rigorous evidence existed on the effects of interprofessional education. It should also be noted that they found no evidence of ineffectiveness. This suggests that the field of interprofessional education is more served by rhetoric than research. It also suggests that this is a fruitful area for more detailed, substantive and sophisticated research.

Interprofessional education is taken to its logical extreme in the UK where there are six courses offering joint nursing and social work qualifications. These joint programmes stem indirectly from the Jay Committee (1979) which was intent on replacing the medical model by a social model in working with people with learning disabilities and following resistance to this view the then General Nursing Council and the Central Council for Education and Training in Social Work came up with the compromise of a dual qualification (GNCs/CCETSW, 1982).

Practice Example

The University of Salford provides a BSc (Hons) joint qualification in social work and learning disability nursing. The impetus for this programme came from learning disability managers who believed that neither qualification suitably prepared staff for employment with people with learning disability. What was needed was a joint qualification. The BSc (Hons) programme at Salford lasts 3 years and is jointly validated by the GSCC and the Nursing

> and Midwifery Council. Students on the programme have to reach the occupational standards of both professions and experience placements in both settings and are taught by both social work and nursing staff.
>
> *Post hoc* surveys have regularly reported that the students have enjoyed their experience of being exposed to two different cultures, learning the ability to work in multidisciplinary teams, and being helped to understand the potential contribution of each profession. On top of this the students also were able to see the importance of holistic assessments and putting the service user at the centre of service provision. When the first students graduated in 1998 they were traditionally employed in nursing although there has been a gradual rebalancing of this position with approximately half now taking up posts in social work or learning disability trusts. Due to the small number of students taking dual qualifications nationally and the lack of independent evaluation it is difficult to estimate the impact of dual qualification holders on practice.

The example above may be seen as either an extreme example of interprofessional education or as an instrument of 'professional engineering'. It is also interesting to speculate on other potential combinations like community psychiatric nursing, health visiting or teaching and social work. Such dual qualifications are not what are meant when we traditionally talk of interprofessional education, but they do offer an alternative way forward.

Interprofessional Practice: The Story So Far

So far we have identified the policy drivers for interprofessional practice, examined the notion of a profession and a discipline and begun to tease out the claimed and potential benefits of interprofessional education and practice. This has also been tempered with some cautionary notes on the difficulties experienced seeking to try and achieve these benefits.

This book is primarily concerned with understanding social work research and the context of practice has been a necessary grounding for our discussion in relation to research. Interdisciplinary research can be understood in at least two different ways – first, researching interprofessional practice and secondly interdisciplinary research of social issues or problems. The first of these has already been identified as an area that would benefit from further research. The second of these is what the remainder of the chapter will be primarily concerned with – examining the nature of interdisciplinary and transdisciplinary research practices.

Interdisciplinary Research

Reflexive Questions

Before going on to discuss interdisciplinary research, what would you consider as interdisciplinary research?

How would interdisciplinary research be different from 'normal' research?

What would you see as the benefits and weaknesses of interdisciplinary research?

Traditional research usually implies research based within one of the major research traditions or disciplines. Obviously it is possible, as has been demonstrated earlier, to consider mixed methods approaches. A multidisciplinary research approach is characterised by the autonomy of the various disciplines, for example sociology or psychology, and does not lead to changes in the existing disciplinary structures. As such multidisciplinary research approaches may help to address a complex mental health problem, but the result of this research will not necessarily impact upon any of the individual research disciplines that have helped to inform the research. Interdisciplinary research, on the other hand, suggests a research approach that is characterised by the explicit formulation of a common framework. Interdisciplinary research does not borrow its research credibility from its constituent disciplines, but creates its own common methodology and may in time become a discipline in its own right.

Within single discipline research centres there is often a statement indicating a desire and wish to work in an interdisciplinary way, which is often translated into working with other single discipline research centres. You might, for instance, have a social work, nursing and an education research centre jointly bidding for safeguarding children's research. Alternatively, some research centres define themselves as a children and families or gerontology research centre, containing a wide range of academic disciplines including social work, health and education and may bid for the same project. There is though a danger in interdisciplinary research that the resulting knowledge production becomes no more than a mere accumulation of knowledge supplied from more than one discipline.

Interdisciplinarity involves the explicit formulation of a single discipline transcending common methodology. This may mean that the form the research takes will involve differing disciplinary researchers working on different themes but within an agreed common methodology. Gibbons et al. (1994) develop this further when they suggest that interdisciplinary research turns into transdisciplinary research when the research is based upon a common framework that is shared by the disciplines involved. The next section identifies the main claims by Gibbons et al. (1994) and their potential impact for social work.

Transdisciplinary Research

> Transdisciplinarity arises only if research is based upon a common theoretical understanding and must be accompanied by a mutual interpenetration of disciplinary epistemologies. Co-operation in this case leads to a clustering of disciplinary rooted problem-solving and creates a transdisciplinary homogenised theory or model pool. (Gibbons et al., 1994: 29)

Gibbons et al. (1994) define two ideal types of knowledge production, Mode 1 and Mode 2. Transdisciplinary research is characterised as the favoured form of research in Mode 2 knowledge production as opposed to Mode 1. Mode 1 is represented as being governed largely by the academic community, mono-disciplinary, homogeneous and hierarchical. Mode 2, in contrast, is described as being carried out at the point of application; transdisciplinary, heterogeneous, more heterarchial and transient whilst its quality control structures are described as more accountable and reflexive. These can be summarised in Table 8.1.

According to Gibbons et al. (1994) Mode 1 is largely governed *by* the academic community *for* the academic community. In this context Mode 1 can be

Table 8.1 *Key characteristics of mode 1 and mode 2 knowledge production*

Key Characteristics	Mode 1	Mode 2
Context governed by	Academic community	Application
Discipline	Single	Transdisciplinary
Nature	Homogeneity	Heterogeneity
Organisational	Hierarchical	Heterarchial
Quality control	Less socially accountable and reflexive	More socially accountable and reflexive
Preferred research style	Single researcher	Collaborating researchers from different disciplines

identified with 'pure science' whose primary goal is the pursuit of knowledge governed by the cognitive and social norms that direct basic research. In contrast, Mode 2 is carried out at the point of application, and is intended to be useful to someone and may be considered as applied science. Earlier in the book we noted that social work is an applied activity whereby the social worker's role is not merely to observe the world but to engage with it in ways to address individual, family, community or social difficulties and problems. Social work research may, and hopefully does, produce new knowledge, but this is not the primary purpose of the research. The primary purpose of social work research is to promote and develop effective social work practice. Thus social work on this criterion can be seen to be a potential Mode 2 discipline. A Mode 2 discipline is organised towards a practical goal and faces the double challenge of being of high academic quality and of practical relevance. This feature of Mode 2 is also transferable to research focused on the 'wicked problems' that require not only the combined and collaborative professions but also the combined and collaborative approaches of differing research traditions and disciplines.

A major feature of Gibbons et al.'s (1994) model is the idea of transdisciplinarity. Transdisciplinarity has four key features:

- A distinct and developing framework guiding problem solving efforts.
- Problem solutions will contain both empirical and theoretical components.
- Dissemination of results is initially accomplished through the process of their production.
- The whole process is dynamic.

Transdisciplinarity refers to a:

> movement beyond disciplinary structures in the constitution of the intellectual agenda, in the manner in which resources are deployed, and in the ways in which research is organised, results communicated and the outcome evaluated. (Gibbons et al., 1994: 27)

In transdisciplinary contexts the traditional boundaries between pure and applied research, theory and practice become less important as the focus is primarily upon the identified problem. Preference is given to collaborative work as opposed to individual performance and excellence is judged by a sustained ability to contribute in the new, temporary and open organisation in which individuals work whilst addressing the issues and problems. Such an approach represents a shift away from fundamental truths towards modes of enquiry oriented towards contextualised results.

Transdisciplinarity favours the diffusion of research results through their process of production and as the research team members move onto new projects. Mode 1, on the other hand, favours communication of results through traditional academic channels of peer-reviewed journals and conferences.

The last aspect of transdisciplinarity is its dynamic character whereby the solution of one problem is the springboard to further advances. Where this knowledge will next be used, or how it will be developed, is difficult to predict. The new knowledge will not necessarily fit into traditional disciplinary boundaries and may help inspire problem solving in different disciplines. Whereas Mode 1 is likely to produce knowledge that will be developed within its own discipline this is not the case for Mode 2, as it is not discipline bound from the outset and can generate knowledge in a range of different disciplines.

The dynamics of knowledge production in Modes 1 and 2 are different with Mode 1 favouring homogeneous growth. An example of this would be the growth in papers in relation to safeguarding children, or evidence-based practice, where the rate of growth often follows a logarithmic curve. This growth essentially consists of more of the same whereby the subject area becomes saturated with increasing numbers of researchers writing about the same themes and issues. On the other hand, heterogeneous growth is marked by the skills and experience of those working on an issue or problem, as these will change and vary over time. Heterogeneity also recognises that there are an increasing number of potential sites for knowledge production, not just unversities but also research centres, government agencies, consultancies, social work employers and practitioners. With the progress in electronic communication it is possible to link these sites as functioning communication networks. Coupled with this there is a growing differentiation at these sites where subject matters are divided into finer and finer specialisms. The reconfiguration of these finer specialisms provides the basis for new forms of knowledge, which over time increasingly move away from the traditional disciplinary boundaries.

One of the key differentiating characteristics between Mode 1 and Mode 2 is the nature of quality control or what is to count as knowledge. Those claiming to produce knowledge have to follow certain rules, but also have to be trained in the appropriate procedures and techniques. Mode 1 is governed by a system of peer review that promotes a system of professionalisation and institutionalisation channelling researchers to work on key problems for the advancement of the discipline and suggesting likely solutions. By comparison Mode 2 is less self-sealing and is likely to create more problematic claims to knowledge, as the knowledge will fall outside the normal disciplinary structures and is therefore less likely to be accepted. Mode 2 not only has to meet the normal academic criterion of acceptance but will also address wider social questions including, 'Will this solution empower or disempower service users?', 'Is this solution ethical?' and ' Who will benefit and who will lose out from this proposed solution?' As such the quality control in Mode 2 is multidimensional bringing to bear a wider range of criteria which does not automatically imply an inferior standard.

In Mode 1 control is exercised by differing knowledge producing institutions, for example universities, whereas Mode 2 occurs within the transient contexts of application and is therefore less likely to receive institutional or disciplinary support. As such Mode 1 could be seen as research defined by disciplinary peers. In contrast, Mode 2 does not have a ready-made group of disciplinary peers and requires a different set of criteria probably including efficacy in terms of the work's ability to provide solutions to transdisciplinary problems. Conventional wisdom suggests that application should follow discovery, whilst for Mode 2 application and discovery are interlinked whereby both happen at once in an iterative process.

Lastly, Mode 1 is characterised by the single disciplinary researcher whereas Mode 2 is characterised by collaborative researchers from different disciplines coming together to address a particular issue or problem. Mode 2 knowledge production can thus be seen as a different form of knowledge production supplementing traditional Mode 1 symbolised by the constant to and fro between the pure and the applied, the theoretical and the practical. Although Mode 1 and Mode 2 have been characterised here as differing paradigms they do interact with each other. Researchers trained in Mode 1 do become Mode 2 researchers. Some of the outputs from Mode 2 may re-enter Mode 1 and act as the springboards for future discovery within that discipline.

Whilst Mode 2 would appear a highly attractive model of knowledge production for social work, it is though not without its difficulties. Gibbons et al. (1994) recognise that such a model is very difficult to achieve and warn against researchers who claim that they are behaving in transdisciplinary or interdisciplinary ways but in reality are disciplinary bound. Different disciplines are identified by their different methodologies, epistemologies and ways of understanding the world that may well be incommensurate. Sociology focuses on the group and psychology on the individual and it is not possible to aggregate psychological understandings to create sociology or to disaggregate sociological understandings to create psychology. Difficulties are likely to be experienced in creating a common language and although it is possible to gain a range of differing perspectives on the same problem the best that may be hoped for is for one of the disciplines to integrate the others. This may include a series of touching points between the different disciplines. It could also be argued that interdisciplinary and transdisciplinary research if successful may become disciplines in their own right. To argue that a multidisciplinary, interdisciplinary or transdisciplinary approach is difficult is different from arguing that it is impossible or an ideal to be strived for. Also, if we are not able to escape the received disciplines how would it be possible to create new disciplines, or how do we explain where disciplines came from in the first place?

Multidisciplinarity, Interdisciplinarity, Transdisciplinarity and Social Work

It is apparent from the above that social work, either by choice or external mandate, cannot ignore the importance of interprofessional approaches in practice and research. To ensure that this does not develop in an *ad hoc* or indiscriminate way requires the development of a robust research culture that is able to address these challenges. To achieve this we have identified three distinct models.

The first of these, multidisciplinarity is represented by a team of researchers who come from either different disciplines or professional groups to focus on the same problem. The different professions may include social workers, health personnel, psychologists and teachers when examining the issue of the educational achievement of children looked after or researchers from a social policy, psychological and sociological perspectives to examine the same phenomena. At the end of this exercise none of the individual professions or disciplines will be changed by the research, as it will be absorbed within each single profession or discipline. Whilst this represents a defensible research position and will be appropriate in many situations it is not appropriate in all. For example, if we accept that the different disciplines of social policy, sociology and psychology all have potential insights into the educational achievement of children looked after, how do we decide which discipline's perspective is the most important and most influential? Without any form of an integrating super-discipline each of the differing discipline's will end up making claims that will be justifiable within their own disciplinary perspective, but incommensurable with each other. For example, if we find that the educational difficulties of children looked after are the result of personal issues, the range of measures to address these will be different than if we discovered that the social policies and structures of health, social work and education were conflictual and undermined looked after children's education.

Interdisciplinarity represents a development of multidisciplinarity whereby the issues of disciplinary primacy become resolved through the creation of a common framework and common methodology. Thus in the example above there would be an agreed process for establishing the different weights of the differing explanations and an agreed way of looking at how and when they individually contributed and what the outcomes were when they interacted. Thus, if it was found that there were personal issues for young people looked after and the organisation and policies of health, social services and education were undermining educational achievement, attempts would be made to understand how these different explanations impacted on each other as well as providing explanations in their own right. Given time the methodology for such a process could potentially herald the beginning of a new discipline.

The third model, transdisciplinarity, represents a progression of the interdisciplinary typology highlighting a model that is focused upon application, heterogeneity and social accountability with collaborative researchers. This model is not restrained by disciplinary boundaries but moves beyond them into new forms of research processes, knowledge production and potential ways of working. In the example of educational achievement such an approach may challenge what is meant by educational achievement and whether young people looked after share this. A transdisciplinary approach with its emphasis upon new configurations of problem solution might also contain organisational, mental health and nutritional specialists as it sought to find effective ways to improve the educational experience and outcomes for looked after children. Such an approach is likely to challenge traditional gaps in terminology, approach and methodology with an emphasis upon developing ways of addressing educational achievement that not only include personal issues, social and structural issues but also seek to combine these in a locally contextualised and applied way. It remains difficult to describe a transdisciplinary approach, but for our purposes it is probably best considered as an extension of an interdisciplinary approach.

Summary

The chapter began by discussing whether social work was a profession and whether it was a discipline. Neither answer was straightforward. The chapter then sought to identify the trend away from individual professional working to multiprofessional working to interprofessional working to address society's 'wicked problems' and to ensure that the service user is at the centre of the service and not the professional. If this trend continues, as seems likely, social workers are more likely to be located in interprofessional teams than in single professional groups. This trend towards interprofessionalism raises important questions for social work including what is the nature of social work and the social work profession, the degree by which social work should be self-policing and the relationship between the state and the professions (Trevillion and Bedford, 2003).

The chapter then discussed the nature and rhetoric of interprofessional education and the benefits claimed for it. Again the evidence for such claims has yet to be robustly developed. This is against a background of increasing uncritical acceptance that interprofessional education is the way forward.

In seeking to promote interdisciplinary working we then looked at multidisciplinary, interdisciplinary and transdisciplinary research. Disciplinary approaches have served us well in the past but are only able to provide solutions within their own discipline and are restricted by their own methodologies and insights. Again it was noted that the new approaches promised much and may potentially deliver much in the future, but have yet to do so. This is not to decry the advances in nanotechnology or the geonome project but to say that in social work research and practice the benefits are yet to be visible.

Suggested Reading

Gibbons, M., Limoges, C., Nowotny, H., Schwartzman, S., Scott, P. and Trow, M. (1994) *The New Production of Knowledge: The Dynamics of Science and Research in Contemporary Societies.* London: Sage. A challenging book which sets out the argument for Mode 2 and the emergence of transdisciplines in science and research.

Journal of Interprofessional Care – this journal is dedicated to researching and discussing issues in relation to interprofessional working.

Leathard, A. (ed.) (2003) *Interprofessional Collaboration: From Policy to Practice in Health and Social Care.* Hove: Brunner-Routledge. This book provides an edited collection of the key issues involved in interprofessional policy, practice and education.

9 Getting Research into Practice

The title of this chapter comes from an initiative in the early 1980s when I was a team leader in practice. Along with a colleague, who was similarly favourably disposed towards research we decided to establish 'GRIP', Getting Research into Practice. (Not to be mistaken with RiP which appeared a number of years later.) Having recruited some similar minded colleagues we organised a conference, which was a great success, and then set up a programme of research seminars for staff in the local authority. These were initially well attended but as time progressed support and energy dwindled with the initiative dying out. Although we tried to resurrect it at a later date this was never successfully achieved. At the stage of the death of the initiative I wondered why my colleagues could not see that research would provide us with the answers to many of our questions and help us to become more effective practitioners and able to deliver a higher quality service.

On a similar theme it needs to be remembered that even the best research does not necessarily change anything. At other times a small study can have effects beyond what could have seemed possible at the outset. It is not possible at the proposal stage to accurately predict the influence that any research may have, although it is possible to try and create a culture in which research messages may be heard.

This chapter seeks to identify why research has not enthused practitioners and why it has not had the impact researchers would have liked upon practice. As part of this discussion we will examine the proclivities of researchers and practitioners/managers, barriers and drivers for research and the nature of research influence through practitioner researchers, inspection regimes and whole organisation initiatives. The chapter will also consider the publicising of research and identifying some of the current initiatives promoting dissemination.

Reflexive Questions

Before we discuss the barriers to research informing practice you should consider whether you agree with the assertion that research has not impacted sufficiently upon practice or not.

What reasons did you give for your answer?

I'd now like you to think about and write down what you see as the barriers to research informing practice.

Barriers to Research Informing Practice

It is always illuminating to ask social work and professionally qualified research students why they do not use research more in their practice. Below is a list of some of the answers. No one has yet suggested that they have maximised the con tribution that research can make to their practice although they are not always clear what has been missing. The barriers from a practitioner view include:

- I'm too busy.
- Relevant research is not easily found.
- 'Lies, damn lies and statistics'.
- If you are caught reading research someone will assume you do not have sufficient work to do and give you another case.
- I don't know where to look.
- You can find research to prove anything.
- Research doesn't address the problems I face in my day-to-day work.
- When you find research it is not easily readable.
- Research complicates matters when all I want to know is what to do next.
- I don't know how to interpret and assess research.

You may have identified some of the above or other possible reasons, but I am sure you will agree that there are many reasons given why social workers do not use research more effectively. It should be noted that this chapter does not set out to blame social workers, or for that matter researchers for this set of affairs. Instead, this chapter seeks to understand the range of factors, including organisational and researcher orientated factors, which militate against a more effective use of research. In many ways this chapter seeks to find a way forward in which research can become 'useful' (for practitioners) and have 'greater impact' (for both researchers and service users). This resonates with our earlier discussions about the quality of social work research achieving the twin standards of practical relevance and academic rigour.

Social Worker Issues

'Too Busy'

From the above set of barriers a number of key issues emerge from the perspective of the social work practitioner. Sheldon and Chivers (2000) have highlighted being 'too busy' or 'pressure of work' as the key obstacle for research informed practice. From 1,341 responses (58.7 per cent of total distributed) pressure of work was the highest recorded obstacle to developing an evidence-based practice approach (mentioned by 98.3 per cent of respondents). This result (98 per cent) was repeated in a follow-up study in 2002 (Sheldon et al. 2005).

This result is not surprising and given the acknowledged national staffing shortage in children and families workers many teams are under severe pressure. The UK Government has responded to this shortage by introducing a recruitment campaign, lowering of the minimum age for admission onto social work qualifying programmes and importantly social work student bursaries.

Currently, training capacity in relation to vacancy rates suggests that 'demand outstrips supply' (Rice, 2003: 8). It is thus likely that there will be a continued demand for social workers with a greater number of posts than personnel. It is also worth reflecting whether even if there were a full complement of social work staff, issues of being 'too busy' and 'work pressures' would still be given as key reasons for not using research? Without a change in the social work culture it is highly likely that this will continue to be the case. For many workers research continues to be viewed as a luxury or add on, rather than embedded in the way social workers think and work. This view suggests that research will only be considered when there is space and will always be vulnerable as not being a 'real' part of the job.

Inadequate Valuing of Research

The importance of research is also reflected in the comment whereby if someone is seen as having time to read a book or journal they are not making good use of their time. Associated with this view is the accusatory presumption that such individuals must also have sufficient space to take on extra work. Thus to be seen reading a journal or book implies you have insufficient 'real' work to do. The perspective represented here is one where research is not seen as a legitimate social work task to undertake in the office. It is also a comment on a view of a social work office as a busy, bustling place where people do not have sufficient time to engage in anything that is not specifically related to practice. Such a view, although understandable, is potentially very dangerous as it neglects the need to continually update knowledge and to critically reflect upon its application to practice.

Between 1998 and 2002 Sheldon and colleagues indicated that 48 per cent and 46 per cent respectively of respondents indicated that they had read literature pertinent to their work in the week prior to the survey with this increasing from 45 per cent to 62 per cent of this occurring at home. In both surveys the most popular publication read was *Community Care. Community Care* is a trade magazine which is free to many staff and contains social work news, opinions, short research pieces, but most importantly is a prime source of adverts for social care posts.

Research Complicates the World

Research complicates matters is, of course, a valid comment in that research and social work have two differing objectives. Research tends to complicate issues by making them more complex to reflect reality whilst practitioners and policy makers are more concerned with simplifying things so that action becomes possible (Booth, 1988). There is a genuine tension here, social workers have to act upon the world whereas researchers help us to understand the world and identify potential interventions. In reality the two operations are linked, but in practice the social worker will not thank the researcher who points out the negative impact of government policies whilst they are faced with a single parent family living in poverty. Both perspectives are required as the changing of government policies may in the longer term reduce poverty but it does not address the immediate, the here and now needs of single parents living in poverty.

In a similar vein it can be argued that researchers are more concerned with emphasising dispassionate precision whilst social workers are embroiled in a world of feelings and messiness. It could be said that researchers are more concerned with the word and social workers with the deed. Researchers become attached to the 'why' question whilst social workers are more concerned with the 'how' question. Stuart (1985) summed up this position very well in relation to managers. If we replace managers with social workers, the quotation also reflects the view of many social workers:

> The stark gap that exists between the dominant providers of research viz academics, and the intended users viz managers. However good the research in academic terms, it does not appear to be a product which well matches the demands of the client. (Stuart, 1985: 20)

Stuart's position for social workers (and for managers) is overstated, but highlights the dilemma by suggesting that social workers and researchers inhabit incommensurate worlds in which researchers and social workers do not interact or learn with and from each other.

One other aspect of this criticism comes from the claim that researchers are more concerned with the development of concepts whilst practitioners and managers are more concerned with their use (Gopineth and Hoffman, 1995). Gopineth and Hoffman (1995) also noted that researchers are able to draw conceptual boundaries around their studies as opposed to the practitioner and manager whose world tends to be less clearly defined and often messier.

Of particular note in this scenario is the growing development of action research, one of whose aims is to simultaneously contribute to basic knowledge in the social sciences whilst at the same time contributing to social action in everyday life (Coghlan and Brannick, 2001). Within this method research acts as an agent of change and the researcher is implicated in the process of change.

Research is Not Valid

There were a few statements we can consider under this heading including: 'lies, damn lies and statistics' and 'you can find research to prove anything'. These two statements suggest that it is impossible to find distinctive answers from research as it can be used to prove anything. We have noted, on more than one occasion, that research is a contested arena, but that this does not mean anything goes. There is also an increasing body of knowledge within social work research about 'what works', where and with whom. Coupled to this there is also an increasing need for a critical research literacy to be able to distinguish between research that should and should not inform practice. Just because research is printed in a research journal or book does not make it valid.

Research Literacy

This section covers the comments concerning the readability of research and the lack of awareness of how to interpret research. The first question identifies a commonly expressed view concerning the accessibility of research evidence.

Traditionally research was contained in long research reports or peer-reviewed journals, written for researchers by researchers. It is thus not surprising that practising social workers found research difficult to access, read, evaluate and identify key practice messages. In more recent times, as we have already noted, the UK Government has sponsored a range of research summaries primarily in children and families work. It is also worth noting that most research reports also include an executive summary that highlights key messages thus giving busy practitioners ready access to the results and recommendations. Also, there has been an electronic revolution with the ability to access research through the world wide Internet, with more and more journals being available on the web accompanied by a growth in electronic databases allowing practitioners and researchers to quickly undertake computer based interrogations of journals, books and research reports at the touch of a button.

This leads us onto the second aspect of the barrier which highlights the difficulty that even if you have the skills to identify and download research articles how will you know the information is valid and reliable? Of course, for peer-reviewed articles these will, as their name suggests, be peer reviewed although this does not guarantee that they will be of the highest quality. The case for books is even less clear as books are written with the intention of making money, not spreading academic knowledge. If a book can disseminate academic knowledge as well as make a profit it is a bonus whilst if it is good science but sells poorly it will be seen to be a failure. Just because something is published in a book does not make it necessarily correct. In order to address this issue of research appraisal the next section examines critical appraisal for social work research before returning to the barriers to research from a researcher perspective.

Critical Appraisal of Social Work Research

The current benchmark statements for social work require social work students to be able to demonstrate 'knowledge and critical appraisal of relevant social research and evaluation methodologies' (Alcock and Williams, 2000: 3.1.4).

As previously noted, for the research-minded social worker it is often difficult to decide which research best meets their needs in deciding how to intervene (or not to intervene) in a particular situation. It is not enough that social workers should merely read journal articles or books. The research-minded social worker must also be able to critically appraise the ever-increasing number of journal articles, research reports and books. Critical appraisal, or critiquing, refers to the process whereby the reader looks for the strengths and weaknesses in a research study. To be able to do this the research-minded social worker has to be 'research literate' and able to differentiate between those studies that they should spend more time on and those they should not. Just because something is published it does not mean that it is worthy of the social workers' most precious resource, their time.

In seeking to critically appraise an article it is worth asking some questions of each stage or section of the research article. This will help to structure your understanding of the subject and help to identify its strengths and weaknesses whilst breaking the task down into a set of achievable tasks. Shepperd (2004) provides a useful starting point when he splits the process of a journal article into problem formulation, literature review, main body of data and conclusion. However, Newman et al. (2005) have developed a more systematised approach to help practitioners appraise quality, decide whether the study is trustworthy enough to justify implementing its results and to consider how relevant the research is to practice. They have developed specialist questionnaires for different types of research study. This book is only going to consider their quasi-experimental and qualitative research tools. Each tool usefully begins with 'screening' questions, which if answered in the negative strongly suggest the reader should not continue working through the rest of the questions and that the reader would not want to apply the research to their practice. If the reader can answer 'yes' to most of the questions it is then likely that the research is of a high quality and that the reader would want to apply the research to their practice or service user decision making. Conversely, if the answer is 'no' to most of the questions then it should probably not be used to inform practice or decision-making.

Reflexive Questions

Find an article in a subject area that you are interested in, using either a quasi-experimental research methodology or a qualitative approach, and apply the appropriate questionnaire, shown below.

Regular applications of these questionnaires will speed up the process and help you to ascertain which articles are deserving of your time and which are not.

Quasi-experimental studies
Please circle the appropriate answers to the questions below

Screening

1 Were the aims of the research clear?

Y N

Please write in the box what the aims of the research were, or what the research question was. (Everything that is done in the piece of research should relate to this question.)

2 Was a quasi-experimental study the right sort of research design to answer the question?

Y N

Did the study aim to find out the effectiveness of an intervention or service, but it would have been very complicated or impossible to have conducted a randomised controlled trial?

Y N

Is it worth continuing?

Y N

Methods

3 Do the researchers justify why a randomised controlled trial would not have been possible or desirable?

Y To some extent N

4 Were the participants recruited to the intervention and control group in a way that minimised bias and confounders?

Y N

5 Were the intervention and control groups well *matched* at the beginning of the study?

Y To some extent N

6 Was the number of participants in the study justified?

Y Not enough detail given N

7 Were the participants in the intervention and control groups followed up, and data collected, in the same way?

Y To some extent N

Results

8 Were *all* participants who started off in the study accounted for at the end?

Y N

9 Are all the results clearly presented and are they appropriate?

Y To some extent N

Briefly describe the main findings

10 Have the researchers taken account of potential confounding factors in the analysis?

Y To some extent N

Conclusions

11 Are the conclusions drawn supported by the study results?

Y To some extent N

Ethics

12 Were the ethical issues addressed?

Y To some extent N

Relevance

13 Is this study relevant to your clients and/or practice?

Y To some extent N

Qualitative Research

Screening

1 Were the aims of the research clear?

Y N

Please write in the box what the aims of the research were, or what the research question was.

2 Was qualitative research the right sort of approach to use or answer the research question?

Y N

Did the researchers aim to interpret or illuminate people's actions and/or their experiences?

Y N

Is it worth continuing?

Y N

Methods

3 Was the research design the most appropriate to address the aims of the study?

Y To some extent N

4 Was the sampling strategy used for the study clear/appropriate?

Y To some extent N

5 Was the number of participants in the study justified?

Y To some extent N

6 Did the way that the data were collected fit with the aims of the study?

Y To some extent N

7 Was the data collection systematic?

Y To some extent N

8 Was it clear how the data were analysed?

Y To some extent N

Results

9 Are the results clearly presented/appropriate?

Y To some extent N

Briefly describe the main results

10 Were steps taken to increase the trustworthiness of the study findings?

Y To some extent N

11 Did the researchers critically examine their role, potential bias and influence on the study?

Y To some extent N

Ethics

12 Were the ethical issues addressed?

Y To some extent N

Relevance

13 Is this study relevant to your clients and/or practice?

Y To some extent N

Both of these appraisal systems have been developed for social work research and provide systematic ways of appraising research material. It should however be noted that an increasing number of research studies use both quantitative and qualitative research methodologies. In these cases social workers could use the appropriate critical appraisal tool for those parts of the research process that are quantitative and qualitative bringing the research answers together at the end. What these tools do not do is remove the requirement from the social worker, or researcher, to engage their critical faculties and exercise their judgement. Critical appraisal tools are not 'black boxes' or technical fixes but structured and standardised ways of making judgements as to the validity and usefulness of a research study.

Researcher Issues

Reflexive Questions

You might like to repeat the first task of this chapter from the perspective of a researcher. What do you consider to be their barriers to research implementation?

In completing the exercise this time you may have considered some of the issues below:

- Researchers do not have the same level of job security.
- Differences in time scales.
- Social workers don't read.
- Social workers aren't interested in research.
- Social work does not have a research culture.
- Researchers are driven by the RAE.
- Researchers want sponsors to have the full copy of their experience and work.
- Researchers don't understand the politics of practice.

Some of the ideas you had may be the reverse of those listed by practitioners, whereby researchers may bemoan the lack of research literacy or willingness of workers to accept complexities and contradictions. However you may be beginning to also consider alternative explanations embedded in the contexts in which social work researchers operate.

Differing Expectations of Research Outcomes

The researcher and practitioner are likely to have different expectations of the research outcomes. Researchers, understandably, favour a substantial report that does justice to all their hard work. Social work practitioners on the other hand favour an executive summary highlighting key issues but most importantly providing 'realistic' and 'real world' practical solutions to their particular problems and issues. The published word is still the standard for judging academic excellence in research. This is also likely to sustain the position that researchers continue to write for research audiences deliberately avoiding non-academic outlets. If this continues to be the case social work researchers will be missing an opportunity to influence practice and will be failing to connect with other audiences restricting the growth and potential of social work research.

Researchers are Politically Naïve

Researchers, managers and practitioners occupy different positions in the political process of the research. It is sometimes claimed that researchers fail to understand the political culture of an organisation and in so doing are unable to fully understand how an organisation operates. Certainly, I have experienced this from both sides of the debate.

Practice Example

As a manager I can still remember a meeting with a researcher who had undertaken research into the implementation of a new policy in our organisation, but whose report the Director wished to change. Prior to the meeting with the researcher the senior management team met together and 'critically' assessed the report looking for points we could 'demand' to be changed with a view of then being able to get the recommendation changed the Director did not like. When the researcher appeared she was warmly welcomed and made to feel comfortable before the full might of the senior management team was brought to bear on her. A couple of years later I met the same researcher in a different context and she advised me that she felt she had been 'beaten up' by the senior management team. She felt that this had been done totally unfairly as her report reflected positively on the organisation, not negatively as presented at the meeting.

From the other side, as a researcher, I have been summoned to a meeting with local authority senior managers where the first item on the agenda was 'what did the director need to know before they met with their business manager?' In this case the politics of the organisation were such that they did not want certain aspects of a particular policy change being described in the way the front-line workers wanted it presented, but instead wanted to present a management perspective on the same policy. Such situations are very difficult as in this case there were two irreconcilable views about the need for change, drivers for change and potential benefits – with the workers blaming the management and the management pointing to agency policies and external events. In this type of situation the researcher can easily get caught up in the politics of the organisation and in so doing offend one group or the other resulting in either the report being left to gain dust on a bookshelf or the workers feeling they have been sold out! Gladly, following some hard bargaining by both sides the report was accepted.

Managers, especially senior ones, spend a lot of their time handling relationships with colleagues, the public and the media seeking to gain and maintain influence. As such, they are tuned to the importance of information and the uses and abuses to which it might be put. It should also be remembered that researchers need to gain access to organisations in order to obtain their data and in this respect managers can be particularly powerful in opening and closing doors.

In these types of situations researchers need to be able to draw upon their value base, interpersonal and political skills to steer a path through such a conflictual situation. Certainly, awareness of the political dynamics of the organisation being researched can help the researcher, but it can also compromise the research if the researcher then seeks to appease the powerful interests within the organisation at the expense of the data being considered. Researchers are not necessarily politically naïve but they do need to be aware that it is the sponsor who has the final say whether any changes result from the research or not.

Researcher Employment

Coupled to the last point is the differing employment positions of the researcher and those researched in relation to the research. The two are often working to different time-scales. This is not simply because managers and practitioners work to short and demanding time-scales whilst good research may take years to complete, write up and publicise. Research contracts will only cover a particular time frame and once the research is completed the researcher needs to move on to another research project whilst organisational managers and practitioners are likely to remain in post.

Differing Worlds of Researchers and Social Workers

It is clear from what has been said that the interplay between the worlds of the social worker and the researcher can, at best, be viewed as problematic. The two worlds suggest a number of incompatibilities in terms of time and time-scales, accessibility of research, differing views of the value of research, differing expectations about the product, and awareness of organisational politics. This can become a defeatist position as it suggests that the received models' view of the researcher and social worker are of occupational groups inhabiting incommensurate worlds. Such a view neglects that both social work and research are essentially mediated activities. Neither activity exists in isolation and both are influenced by internal rationale, values and external factors, which serve to mediate and influence the relationship between research and social work. Research impacts upon practice, for example the work of Parker (1966) on the likelihood of placement breakdown and the age of foster carer's own children has influenced generations of social workers practitioners. On the other hand practice influences research, not only because researchers want to study the issues of the day, but also because at least some of these issues, for example, interprofessional working, reappear as among the priorities of funding bodies.

Organisational Issues

Culture

Within this chapter we have explored some of the barriers for social workers being able to utilise research and for researchers to be able to have their messages heard. Within this process a number of problems have been suggested as social work issues but are in many ways responses to organisational imperatives. The taken for granted assumptions within an organisation, the structures within which social workers undertake their task, mediate and play a part in promoting or hindering the development of a research culture. The way organisations use, support, talk about, commission or not commission research provides subtle messages to staff about the importance given to research and its findings. The very culture of an organisation can promote or inhibit the use of research. Schein has usefully defined an organisational culture as:

> The pattern of basic assumptions that a given group has invented, discovered or developed in learning to cope with its problems of external adaptation and internal integration, that worked well enough to be considered valid and, therefore, to be

> taught to new members as the correct way to perceive, think and feel in relation to those problems. (Schein, 1984: 3)

This classic definition helps us to understand the pervasiveness of culture and how it becomes transmitted to new members 'as the way things are done around here'. Culture acts as a filter, allowing some ideas into the organisation and filtering others out. In relation to research it is possible to see how the culture, often from the top down, but not necessarily so, can promote the development of a research informed culture or act to prevent research becoming part of the organisation's 'way of doing things around here'.

Managerialism

This has been discussed much more fully in **Chapter 1** and readers may wish to revisit this chapter. In terms of social care organisations, performance on key indicators and star ratings have increasingly become major driving forces for change.

Performance indicators can represent both a push towards a more creative practice or a re-entrenchment of safe practice. The Government's indicator on stability of placements is for no more than 16 per cent of children looked after to experience more than three placements in a year. This indicator rightly encourages social workers, and their managers, to find the right placement first time, but it may also stop young people moving onto to more appropriate placements if it was going to impact negatively on the authority's performance indicator. An over concern with performance indicators will lead to a functionalist practice orientation that will inhibit creativity and learning. What works will become defined as what achieves good outcomes in the performance assessment framework. Agencies report that the principle of measuring performance is right, but that the 'how' of both content and delivery are more open to question. This is particularly the case for performance indicators as Law and Janzon (2005) report the jury remains out as to whether their introduction has resulted in better services. Whilst it would be naïve to suggest that social workers and their managers should ignore the performance framework, this is patently absurd. Social workers, and their managers, need to continually adopt a critical attitude towards the indicators or best practice will become subsumed by the achievement of star ratings and performance indicators rather than dealing humanely with the complexity and messiness of the service user's difficulties and problems.

Resources

Most social care organisations will have libraries, access to the Internet and policies documents that are all potential research resources. It is only in recent times that basic grade social workers in many of these organisations have had access to the World Wide Web and Young et al. (2006) identify how such access can be supportive of research and non-access an inhibitor for research. Sheldon et al. (2005) note that in their 2002 survey only 58 per cent of respondents had access to the Internet and a quarter had no access to library facilities. How widely available such opportunities are within an organisation provides subtle messages as to how that organisation encourages or discourages research.

Staff Development

How organisations manage and promote their staff development and training opportunities is also important. If social care organisations only provide training in those matters that are directly related to practice, for example the implementation of a single assessment framework for older people, they are in danger of becoming increasingly functional and rule bound in their practice. Besides the necessary training on policies and procedures staff development opportunities and training events should also include the possibility of learning about wider social care issues informed by the latest research, attendance at workshops run by RiP or MRC or similar organisations. Social workers are expected to demonstrate their continued professional development to re-register with their care council and organisations have an opportunity here to promote activities to support the development of research mindedness. The organisation may also sponsor staff onto post-qualifying courses or higher degrees. All these activities, including any support that is offered in terms of time or finance, will help to develop the research orientation of the organisation's culture and the place of research as a legitimate activity within the organisation.

Developing a Research-Minded Culture

Walter et al. (2004) undertook a knowledge review for SCIE into improving the use of research in practice from which they identified a tri-partite model for thinking about and developing research-informed practice. This model consisted of the research-based practitioner model, the embedded research model and the organisational excellence model. These models are explored in greater detail below alongside the ideas of the practitioner researcher and the learning organisation.

Research-Based Practitioner Model

The research-based practitioner model is based on four key premises:

- It is the role and responsibility of the individual practitioner to keep abreast of research and ensure that it is used to inform day-to-day practice.
- The use of research is a linear process, accessing, appraising and applying research.
- Practitioners have high levels of professional autonomy to change practice based on research.
- Professional education and training are important in enabling research use. (Walter et al., 2004: 25)

This model focuses on the day-to-day experience of the individual practitioner as to how they operationalise research to help inform their practice. The individual practitioner is seen as having the major role in searching out and finding relevant research which they then put into practice as independent professionals. This model is also supported by the DH requirements for social work training that embed the use of research as part of the national occupational standards at pre- and post-qualification levels (GSCC, 2005; Topss, 2002).

Inherent in this model is a linear view of research whereby research findings are accessed and appraised by practitioners who then apply the results to the problem in hand. This model, besides being linear, is uni-directional as the practitioner is seen as the recipient of knowledge and plays no part in the research process or identifying what research should be undertaken.

Walter et al., also report that although there has been a wide range of initiatives and activities to support access to research for social care staff 'they had found no studies that had formally evaluated the success of such initiatives' (2004: 28). The notion of the research based practitioner is currently an act of faith awaiting evidence to identify whether it works, and if so, what conditions can best support its development.

The next section develops the idea of the research-based practitioner further and seeks to examine the potential of practitioner researchers.

The Practitioner Researcher – Contributing to Both Worlds

The practitioner researcher is a special case of the research-based practitioner. You may well know a colleague who is undertaking research either because of a personal interest, for a higher degree or as part of the post-qualifying framework.

You may even be a practitioner researcher yourself or considering becoming one. A practitioner researcher is a social worker that, either in their spare time, or as part of their current employment undertakes research which is small-scale and carried out by the professionals who deliver these self same services. This research is likely to have local relevance and focus on individual services and the critical assessment of research findings to evaluate current provision and develop new, more relevant, services. One potential area of collaboration is for practitioner researchers to be supervised by university research personnel either alone, or potentially more helpfully, as part of a group or a learning set. These arrangements could benefit both the universities and practitioner researchers but require employers to become more imaginative in how they support their staff.

Employer-led support for practitioner research may involve providing time, research resources, access to files or people which might be done for a fixed period or as part of a permanent post. Practitioner research also highlights the tension within social services whereby the employer may act in a range of roles including those of 'benign observer', 'critical friend' or 'controller'.

In the 'benign observer' role the employer allows the researcher to undertake their research with little interest or interference in the subject chosen, the methods deployed, the analysis of the data and any recommendations that flowed from the results. The authority discharges its minimum responsibilities via research governance in response to the Department of Health's Research and Governance framework discussed earlier. This type of organisational practitioner research model is representative of employers who have a laissez-faire attitude towards research and do not value the potential benefits it may bring.

In the 'critical friend' role the employing organisation takes a more proactive role. Here, the employer seeks to promote the research but also wants to ensure its validity and credibility so that its results will be based on valid and credible data. In this scenario the employing organisation goes beyond its minimum duties and seeks to maximise the interests of its staff group and will also wish to support projects that align themselves with the organisation's strategic plan and potentially impact on the way the organisation or its workers operate.

The 'controller employer' is the type of employer who not only dictates the research topic but also the methodology to be applied and how the results will be disseminated (if at all). In this type of organisation the practitioner researcher becomes a tool of the management. Management oversees the research to ensure that the analysis and results will reflect their view on the issue under investigation and that any recommendations will be those that they want.

On first examination the organisation as 'critical friend' seems the most desirable of the three typologies, but is not necessarily the most appropriate in all cases. If the area under study is of limited importance to the organisation and is being supervised by a reputable researcher the organisation may decide that their investment of time will provide minimal benefits. Thus the first two types may be appropriate at different times for the practitioner researcher whilst the third type is much more problematic. In these circumstances it is a mute question whether the practitioner researcher would be better advised not to undertake the research at all given the interference they are likely to receive from their managers.

Limitations of the Practitioner Researcher

As we discussed earlier social workers already have some of the key skills required of a researcher in terms of interviewing and analytical skills. As a member of the organisation they are researching, a politically aware practitioner researcher is more likely to be able to identify whom they need to gain approval from, who would make an effective ally and who is likely to be cynical or disruptive. These are issues that often cause difficulties and take professional researchers a long time to discover. On the other hand practitioner researchers may have a particular difficulty because of being known to their organisation. This can either make access easier or cause colleagues to behave badly if they do not value the worker in the first place. Practitioner researchers also have the difficulty that some of their colleagues may find it difficult to see them outside their role as a colleague and seek to restrict the opportunities for their colleague to behave as a researcher.

Shaw (2004) raises more critical concerns about the role of the practitioner researcher and their limited ability to influence agency and wider social work practice. In particular he highlights issues to do with the research being agency owned, employer led, a solitary activity, non-collaborative, non-collective, insider research and bounded in both professional and discipline terms.

The Embedded Research Model

The embedded research model is based on the view that:

- Research use is achieved by embedding research in the systems and processes of social care, such as standards, policies, procedures and tools.
- Responsibility for ensuring research use lies with policy makers and service delivery managers.
- The use of research is both a linear and instrumental process: research is translated directly into practice change.
- Funding performance management and regulatory regimes are used to encourage the use of research-based guidance and tools. (Walter et al., 2004: 26)

In this model, practitioners are not expected to actively seek out research evidence. Research enters practice via the infrastructures of regulatory frameworks, inspections, policies, procedures and practice tools. In the research practitioner model the key link was between research and practice – in this model it is between research and policy. This model does not assume practitioner autonomy and in fact may seek to restrict it in order to meet the embedded research-based performance management indicators.

This model does have a major advantage over the practitioner model as the difficulties experienced by practitioners in accessing research whilst managing workloads are overcome as this task falls to policy makers. This model does though assume that practitioners do not wish to engage their powers of critique and appraisal and are willing to do what their organisation asks them to do without question. This model does not require practitioners to own the research they use, only to apply it. This creates the danger that the embedded research model is open to becoming another example of management control.

Having examined the practitioner-based model and the embedded research model we now move onto the organisational excellence model.

The Organisational Excellence Model

This model is based on:

- The key to successful use rests with social care delivery organisations: their leadership, management and organisation.
- Research use is supported by developing an organisational culture that is 'research-minded'.
- There is local adaptation of research findings and ongoing learning within organisations.
- Partnerships with local universities and intermediary organisations are used to facilitate both the creation of use and research knowledge. (Walter et al., 2004: 26)

Within this model the responsibility for developing a research-informed practice doesn't lie with the individual practitioner or policy maker, but with the organisation and in particular its leadership and management. This model recognises that individual social workers are constrained and empowered by the structure and culture of the organisations in which they work. In particular this model places organisational learning at the fore. Research knowledge becomes integrated into organisational knowledge and the view of research becomes cyclical rather than linear. Within the model there is a focus on local learning that is often collaborative between researchers and practitioners whereby practitioner knowledge becomes integrated with research knowledge. Practitioners test out research knowledge finding out what works and what does not, informing researchers so that they can consider their findings in relation to the practice experience and further sensitise their research to this new information. This creates an iterative and dynamic example of research supporting practice and practice supporting research. However, the role of service users in this model is unclear and although there is space within the model to include service users there is no mandate to do so.

This organisational excellence model also points towards the current interest in learning organisations, which is explored below.

The Growth of the Learning Organisation

In recent years there has been an increasing interest in the notion of a learning organisation, first in the private sector, and more recently in the public sector.

> **Reflexive Questions**
>
> Think about any team or organisation that you have been a member of. Would you have described it as a learning organisation?
>
> What do you consider are the essential components of a learning organisation? Do you think that such an agency is more likely to happen in the statutory, voluntary or private sector?
>
> Why?

Background to the Learning Organisation

The notion of a learning organisation appears intuitively attractive – who would want to work in a non-learning organisation? Besides being inherently attractive, what does it mean? The concept of the learning organisation appears to have developed from two distinct strands. The first of these is concerned with the nature of organisations and their imperative for environmental adaptability to survive. The second strand developed out of the increasing awareness of the limitations of traditional staff development wherein individuals would be developed but this did not necessarily translate into organisational impact or better quality services. This represents an ongoing struggle for the compatibility between personal growth and organised human relationships. The learning organisation has been informed by the work of both Schon (1971) and Bateson (1972). Schon brought the term 'learning system' into the mainstream of organisation thought and Bateson is recognised for his work on types of learning which was later developed by Argyris and Schon (1978) who suggested most organisational learning was single loop (error-detection and correction) and that only rare glimpses of double loop learning could be found. Double loop learning results in changes to operating assumptions, norms and values. There is also a third type of learning called deutero-learning, which is concerned with learning to learn and occurs when organisational members engage in reflecting and enquiring into previous episodes of organisational learning or failure to learn. In doing so they discover what has facilitated or inhibited learning and are thus able to invent new strategies whilst evaluating and generalising what they have produced. This begins to point towards the characteristics of the learning organisation which cannot be brought about by training individuals – it can only happen as a result of learning at the organisational level.

> A learning company is an organization that facilitates the learning of all its members and continually transforms itself. (Pedler et al., 1991: 1)

This concept serves as a statement of a managerial goal that an organisation can be redesigned so that it is capable of adapting, changing, developing and transforming itself. The learning organisation not only delivers services, for action is not enough. Action, in the learning organisation seeks to resolve the immediate problem but, just as importantly, it also seeks to learn from the process. Action then becomes not merely the search for the 'right answer' but service provision as experiment. Service provision becomes reframed as a form of action research intended to improve effectiveness and performance evaluating practice and learning from both success and failure. Ideas like 'partnership', 'integrated services' and 'single assessment' become experiments rather than solutions and we begin to learn from them rather than being swept along with the latest fad.

Baldwin (2004) also notes that a value base of promoting participation and democracy is essential to support the development of the learning organisation. To avoid becoming no more than a slogan the learning organisation is produced out of social relations including those that characterise inequality including relations based on gender, class, disability, 'race', sexuality and so on (Gould, 2004).

Participation within a learning organisation does not stop with the employees of an organisation but includes all its stakeholders as all are seen to benefit from the increased learning potential. In social work these stakeholders would not only include other social care organisations that commission or deliver services, but all those who can contribute to meeting the needs of the organisation's service users. Stakeholders would also include service users whose knowledge of the impact and appropriateness of social work practice would be valued and systematically collected to inform the organisation of its performance.

In the conclusions to an edited book on learning organisations Baldwin (2004) comments that there is no evidence of fully functioning learning organisations within the social care sector, but there are glimpses appearing in some organisations.

Barratt and Hodson (2006) as part of RiP have identified five key foundations to create an infrastructure for a more research-informed practice organisation that goes some way to becoming a learning organisation. These key foundations are:

- Giving a strategic lead – need for a senior manager to spearhead the initiative.
- Setting expectations – what evidence should be used where, how and by whom.
- Encouraging learning from research – promoting a learning organisation.
- Improving access to research – making it easy for staff to access high quality relevant research.
- Supporting local research – encouraging staff to generate their own evidence during the course of their own practice.

To support their foundations they have developed a practical guide and series of tools to enable social service teams to audit their practice and identify strategies for promoting greater use of research including:

- Developing a vision of evidence-informed practice and communicating it.
- Establishing a steering group with a clear action plan.
- Including importance of research-informed practice in recruitment and selection processes, induction, and supervision and progression criteria.
- Winning the hearts and minds of first-line managers.
- Developing a learning culture.
- Supporting staff to keep themselves up-to-date.
- Developing local research libraries, access to Internet and research updates.
- Enabling service self-evaluation.

In a similar vein, SCIE (2004) has also developed a self-assessment resource pack for developing learning organisations. The UK Government has adopted the notion of a learning organisation as an aspirational goal within its modernisation agenda (Department of Health, 1998, 2001e, 2001f). However, this self same governmental interest in the learning organisation represents a contradiction as Baldwin (2004) views it as a major threat to its realisation. He notes that managerialism and the learning organisation represent mutually distinct ways of conceptualising organisations in terms of how they should be structured, who should work in them and how employees should be treated. In particular he is concerned that the learning organisation as constructed through social relationships is in danger of being procedularised and structured so that:

> It is almost as if this discourse has taken the words of the learning organization development and subverted them in a tokenistic fashion that enables the speaker to sound as if that is the goal while never having to walk down the road towards the destination. (Baldwin, 2004: 164)

Gould (2004) notes that the learning organisation, by definition, is seeking to continuously learn from and influence its changing environment and must itself always be in the process of change.

On initial examination the promotion of a learning organisation suggests a supportive environment for the research-minded practitioner. On closer examination the promotion of this concept to support the Government's agenda may militate against its development or represent a form of colonisation. Workers and students need to be concerned that the promotion of the learning organisation does not lead to the development of the more routinised 'evidence-based practices' (see **Chapter 5**), restricting creativity and whilst promoting instrumental views of the social work task ignoring the milieu, values and contradictions in which social workers are asked to perform.

Improving the Use of Research for Practice

We have examined different ways that research might be introduced to practice via two practitioner models, a model of embedded research and two organisational models. None of these models has all the answers. The practitioner models focus on individual responsibility and autonomy of practitioners to impact on their individual caseloads or service area. The embedded model with its emphasis upon policy requires the worker to be the instrument of policy, implementing policy and procedures within regulatory frameworks over which they have little influence. The organisational models emphasise the importance of organisational learning and that individual learning is not enough. The learning company also emphasises the importance of participation and how stakeholders can contribute to a dynamic model of research, policy and practice.

In their analysis Walter et al. (2004) argue for a whole systems approach as the way forward for social work. This approach acknowledges that each of the models are limited but that together they begin to impact upon all the different aspects of the social care system. Whilst there is considerable value in such a suggestion there are significant tensions between the models: the practitioner model's emphasis on individual autonomy and responsibility is at odds with the embedded research model's instrumental view. This instrumental view is in conflict with the excellence and learning organisational models that emphasise collaborative and participative learning, and a joint iterative research learning process that is opposed to the non-reflective linear policy research process in the embedded model. Whilst conceptually it is clear that there are tensions between the different models this is not to say that organisations are incapable of holding the tensions and contradictory aspects of each of the models in their own service delivery processes. The emphasis on regulation and accountability will ensure the embedded model is part of all social care organisations. In the areas where there is choice the research-based practitioner and practitioner researcher can co-exist within the organisational excellence and learning organisation formats as long as the individual learning does not get stuck at the level of the individual and processes are established to share this knowledge. It is also true that different problems will require different research approaches and what may be relevant for one problem will not be relevant for another or at different points of the same research cycle. What is needed though is some research evidence to identify under what circumstances and for what problems these models can be combined effectively. On a purely pragmatic level Sheldon et al.'s (2005) research found that social workers said they could become more research minded if they had increased access to technical research facilities, protected study time and opportunities to attend research meetings. To this could also be added research that is accessible, understandable, answers practice questions, that will benefit service users, is supported by social work organisations, that will enhance social workers' professional status and workers will be rewarded for using research (Newman and McDaniel reported in Newman et al., 2005: 140).

Summary

This chapter has sought to begin to identify issues concerned with why research has failed to make the impact upon practice that researchers would desire. We focused on three major areas of concern linked to social workers, researchers and organisations. In relation to social workers we highlighted problems concerning social workers being 'too busy', not valuing research, research complicating matters and a lack of research literacy amongst social workers. These problems were then considered further when the researcher perspective was highlighted noting the differing expectations for research outcomes of researchers and social workers, the political naïvety of researchers, the employment position of researchers and the differing worlds of social worker and researcher. The analysis then moved onto how organisations in terms of their organisational cultures, drivers, and training and staff development can promote or inhibit a research-minded practice.

Following this we then looked at ways of getting research into practice and bridging these tensions. In particular we identified the importance of practitioner models, the embedded research model and organisational models. All three models' strengths and weaknesses were identified. It was then suggested that together all three approaches can help promote a holistic research informed practice but that we cannot totally ignore the tensions between the different models. Lastly, it was also noted that we do not have any robust evidence that these approaches will work and more research needs to be focused on promoting research in practice.

Suggested Reading

Barratt, M. and Hodson, R. (2006) *Firm Foundations: A Practical Guide to Organisational Support for the Use of Research Evidence.* Dartington: Research in Practice. Provides a rationale and practical exercises and support materials to develop a research-informed practice.

Gould, N. and Baldwin, M. (eds) (2004) *Social Work, Critical Reflection and the Learning Organization.* Aldershot: Ashgate. This is a useful book for those wanting to begin to understand the nature of learning organisations and their potential applicability to social work.

Sheldon, B. and Chivers, R. (2000) *Evidence-based Social Care: A Study of Prospects and Problems.* Lyme Regis: Russell House Publishing. A useful discussion on evidence-based practice and the research evidence on the barriers to social workers' willingness to use research.

Walter, I., Nutley, S., Percy-Smith, J., McNeish, D. and Frost, S. (2004) *Knowledge Review 7: Improving the Use of Research in Social Care Practice.* London: SCIE. A comprehensive review of the evidence of research being used to inform practice. This publication is available free from SCIE.

10 Whither Social Work Research – Challenges for the Third Millennium

This chapter begins with a review of the previous chapters looking at the implications of what we have discussed so far for the development of the research-minded social worker. Following this analysis we look toward potential futures for social work and the role that research might play in that future. This discussion is based within an acceptance that the notion of social work is contested and changing, but it is this uncertainty that makes it all the more important that researchers and social workers enter into a dialogue. This dialogue must begin to explore how research can help to inform practice and how practice can act both as a springboard for research and as a critical friend promoting further dialogue, an increased sophistication and sensitivity of research evidence. This suggests a model of research and practice that is not linear, but iterative where research and practice are continuously critiqued, one by the other, helping each to become more specific, useful and relevant.

At the beginning of the book it was claimed that good research can be synonymous with good practice and the book has sought to show how and why this statement should become a reality. Preston-Shoot (2002) has lamented that social workers have all too often been expected to pay little attention to theory, research and the literature underpinning their profession once they have become qualified. On entering employment the practice culture of their new colleagues and supervisors takes over and what matters most is 'moving the work' rather than working with the service user to address identified need. It is still not unusual to hear newly qualified social workers being told that they are in the 'real world' now and they can forget everything they learned on their course! Such a view begs the question why such organisations need qualified staff. It also questions whether students should join such organisations that appear only to want those who will do what they are told, will not question why, and are not encouraged to contribute to the organisation using their experience and knowledge of research to develop better practice. Before moving onto potential scenarios for the future of social work research it is helpful to be reminded of some of the key debates and discussions as identified in this book.

Reflexive Questions

Before the key themes are reviewed you might like to put down four bullet points of what you consider to be the most important learning points or themes you discovered from reading this book.

Back in **Chapter** 1 we identified why research was critical to the continued development and survival of social work. This was against a backdrop in which the nature of social work and, by definition social work research, is contested and changing. As society changes, the relationship between those who provide and receive social work services changes. The book identified the agreed IASSW and IFSW (2004) working definition which emphasises the importance of the social worker utilising theories of human behaviour and social systems to promote change in human relationships, promoting the well-being, empowerment and liberation of people. This definition highlights the key importance of human rights and social justice as fundamental to social work and its practice. This begins to identify the research agenda for social work and the values and interests that drive social work research.

This acceptance that social work research is grounded in the 'real world' also helps us to begin to understand that the nature of social work is mediated by the nature of society. In this context social work practice is contextualised by the nature of society we inhabit in this the third millennium, the legal system in which it operates and also the sites where it is practised. It is against this context that this book has sought to understand and explore the importance of research for social work practice and the research business.

The Research Business

One clear message from this book is that social work research is rarely undertaken solely for the purpose of enhancing and developing our knowledge. Social work is a practice discipline, and it is incumbent upon those who see themselves as social work researchers to consider how their research can be translated into behaviours to actively support practice. This does not mean that pure research is eschewed but that it is used to promote understanding and to act as a springboard for applied research and practice. Of particular concern is the underfunding of social work research in comparison to health research and the likely impact of this disparity on the infrastructure of social work research and modernisation of social care. The principle measure for the quality of research is published research papers in peer-reviewed journals – by definition this writing is for an audience of academic colleagues not for social work managers or practitioners.

Research Philosophies

Having identified the business of social work research and its importance to the development of social work practice we then moved on to examine two of the major paradigms of social research, positivism and interpretivism. The underpinning philosophies including ontological and methodological were identified and critiqued. It was noted that positivism was linked to quantitative methods favoured in the USA and interpretivism to qualitative methods favoured in the UK and Australia. Both methods were seen to represent mutually incommensurate points of view. To believe in one is by definition to disbelieve in the other. This philosophical impasse is circumvented in practice whereby researchers often use mixed methods. Either the researchers are deluded or the philosophical incommensurability is only that, philosophical. A mixed methods approach does not though sidestep philosophical issues by mixing the methods and researchers need to be aware that merely adopting a 'chocolate box' approach to choosing methods is not necessarily appropriate either.

Social Work Research is Ethically Driven

Chapter 4 examined the need for ethical research highlighting some of the abuses that had occurred to research subjects in the past. This chapter also sought to argue that ethical research was much more than ethical approval from a research governance and ethics committee. This was not to say that gaining such approval is not important, but that often the most difficult of ethical questions occur once the research has started and where not envisaged in the original approval. To this end it was argued that social work researchers need to become morally active throughout their research study and examples were provided as to some of the ethical issues that might be faced at different points in the research process.

Throughout the book the reader is reminded that to act ethically in research is not merely desirable but essential. Good intentions are not enough – the social work researcher has to be able to show that they have considered as fully as possible any likely harm their research may present to research respondents and how these risks can be reduced. Ethics are inherently embedded in social work practice and must also be similarly embedded in social work research.

Evidence-based Practice – Today's Success Story

Having previously examined the importance of philosophical approaches we then moved onto the growing importance of evidence-based practice. Evidence-based practice is strongly supported by governments, as it is seen to promise effective economical interventions. In trying to understand evidence-based practice it was necessary to return to medicine where it first became established and has made substantive strides. Evidence-based practice was seen to be based on the conscientious, explicit and judicious use of current best evidence, all terms which are open to challenge and debate. Evidence-based practice steer clear of this debate and instead looks towards a hierarchy of research evidence with the systematic reviews of randomised control trials at the apex and expert consensus and individual opinion at the bottom. This hierarchy of evidence is heavily weighted towards behaviourist and positivistic assumptions involving quantitative methods. Whilst these assumptions may hold true in medicine (although even here there are questions being asked) it is questioned whether the same assumptions also hold true for social work. Evidence-based practice presents a rational-logical approach that does not necessarily transfer well to social work. In particular social work does not have a sufficiently developed evidence base and even if this was possible the degree of certainty unlike biological knowledge poses inherent problems in an uncertain and complex world. It was also noted that social work researchers had welcomed the attention that evidence-based practice had brought to the research contribution to practice. Social work researchers have adapted this interest in research to talk about research informed practice emphasising a distancing from the hierarchy of knowledge and the worst of the logical rational assumptions.

Service Users: The New Research Orthodoxy

One of the major changes in recent years has been the centrality given to service users within the world of social work. As previously indicated the modernisation agenda has stressed the importance of placing the service user at the centre of service provision.

In recent times there has also been an increasing interest in involving service users in research that is seen as both beneficial to service users and to the research. Service users may benefit the research by offering a different perspective, ensuring the issues identified are important to service users, ensuring the questions are relevant and understandable, aiding the recruitment of potential research respondents and promoting the dissemination of results. Nonetheless researchers remain reticent to involve service users as co-researchers citing difficulties about resources, effort and service user representativeness.

However, if service users are to be involved in research it is necessary to consider what aspects of the research they should participate in, to what degree and what training they will require to be able to discharge this task effectively. As yet we are only in the process of exploring the limits of involvement. It is unlikely that service users will have all the skills of a researcher, which is not to say that with training and practice many of these could not be taught. It is though open to question whether a point is reached when the service user becomes more researcher than service user. This becomes partially paradoxical, as it is the service user's experience that is valued and it is this experience that contributes to new understandings in the social work research and practice processes. It also needs to be remembered that involving service users in research does not necessarily mean better research or outcomes. Service user research is no panacea but an alternative research approach with its own strengths and weaknesses many of which we are only now coming to understand.

Anti-oppressive Research

Building on from the importance given to the service user perspective on research is the need to develop anti-oppressive research practices. Social work is unequivocally committed to anti-discriminatory and anti-oppressive practice. Both concepts are often conflated one into the other. What anti-discriminatory and anti-oppressive research practice actually is though is historically highly contentious and open to debate and conjecture. Concepts like 'blackness' and 'whiteness' have been shown to be problematic and potentially restrictive. Anti-oppressive practice in research though does contain a commitment to highlight, challenge and eradicate social injustice at whatever levels within society such oppression or discrimination should occur. For the social work researcher this represents a double challenge in that the research process itself should not be discriminatory or oppressive and that the areas of research focused upon should represent a commitment to an anti-oppressive stance. This can create difficulties in gaining access to research subjects as which employer is likely to agree to a research project that shows themselves or their staff as oppressive or discriminatory?

Traditionally anti-oppressive practice is associated with issues of race, gender, disability, sexuality and age. More recently issues of colonialism and urbanism have been added to this list with an acceptance that there may yet be more areas of anti-oppressive structures that have yet to be discovered. It also needs to be acknowledged that individuals are not uni-dimensional, but have multiple cross-cutting dimensions in that they may be a black, female, paraplegic, homosexual, older person of Maori descent living in a rural setting or any other combination of the differing discriminatory identities. It is also possible that any individual will both be the subject and perpetrator of oppression in different aspects of their lives. Social work researchers need to be aware of the danger of falling into an uncritical stance in response to anti-oppressive research and the hierarchy of oppression trap where it is claimed that one aspect of oppression is more oppressive than another. The current conflicts over immigration, refugees and asylum seekers bring these issues to a head challenging both the social work researcher and social worker as to the nature of their role in this highly politicised arena.

Interdisciplinary Research and the Research Process

In recent years there has been an increasing emphasis upon the interdisciplinary contribution to social work and social work research. Recently we have seen the ending of social service departments and the development of a new mosaic of social work sites of practice. In many of these sites social workers are required to work with other professionals in the same service. In order to address society's 'wicked problems' it has become accepted that no one profession has the skills or the abilities to do it all by themselves. The new orthodoxy emphasises the importance of joined-up-thinking and joined-up-working. This has transferred to the research process and it was questioned whether social work could be seen as a discipline as it relies heavily upon knowledge domains including sociology, education, psychology, law, social policy, philosophy and economics.

For social work research the nature of interdisciplinarity and interprofessionalism poses two particular challenges. The first of these is to identify under what conditions interprofessional working achieves better outcomes for service users and to identify ways in which interdisciplinary research can become an effective research tool. **Chapter 8** also discussed the nature of Mode 2 knowledge production and how this appeared to be more reflective of much current social work research with its emphasis upon application, transdisciplinarity, heterogeneity, heterarchial, social accountability and collaborating researchers from different disciplines.

Getting a Grip on Research

Chapter 9 sought to identify how practice can be used to inform research and vice versa. In particular we highlighted the barriers to research from a social worker's perspective, researcher's perspective and from an organisational perspective. From this it was clear that there were many barriers to the successful incorporation of research into practice. These barriers included social workers being 'too busy', not valuing research and research not being in an accessible form. For researchers the barriers included issues in relation to the RAE, differing expectations about research outcomes and the format in which research should be published to promote researcher careers. Organisational culture and managerialism with its focus on performance indicators can inhibit or promote research within an organisation. One way of beginning to (re)align these differing interests between social work researchers, social workers and organisations was the notion of the learning organisation. The learning organisation encourages reflective and reflexive practices whilst seeking to continually facilitate the learning of all its members and at the same time emphasising interconnectednesss and connectivity. Within the learning organisation action is viewed not as an attempt to resolve a problem but is also seen as a learning opportunity where service commissioning and provision are seen as experiments to learn from.

Walter et al. (2004), in a review aimed at improving the use of research in practice, also highlighted three other ways in which research can inform practice in what they termed the practitioner research model, the embedded research model and the organisational excellence model. For Walter et al. (2004) it was not a case of whether one model is better than any other but a need for all three models to be used to promote a holistic strategy for the promotion of research to inform practice. Such a view though does not say anything about how these three strands can be balanced and whether too much of one will negatively impact upon the others.

Messages from Social Work Research

Research like social work is a messy social and political activity. As we have seen in earlier chapters it is highly contested, fraught with ideological, ethical and moral debates. As Humphries notes:

> Neither social work nor social work research is a neutral activity. Like other social practices they are both subject to external and internal forces that influence them towards being instruments of 'empowerment' or otherwise. That they are unambiguously a 'good' cannot be taken for granted. (Humphries, 2005: 279)

Social work exists in those grey areas within society where it has to grapple with the tensions inherent in providing care for those who are unable to care for themselves and to protect those who are unable to protect themselves. Similarly social work inhabits the intersections between service users and carers, balanced between social work managers and other social workers. Given this uncomfortable position one beacon to help social workers steer a course through the competing demands and interests is to be able to use social work research that has been informed by the same value base as social work itself. Such a knowledge base will seek to empower, challenge oppression and engage service users as human beings.

The Future of Social Work Research

At this point it could be very easy to be pessimistic about the future of social work research and social work practice. As we have described in this book, the UK is currently in a period of contradiction where a myriad of social work practice sites have replaced the old social work structures heralded by the Seehbohm Report (1968). The very terms social care, social work and social work research are contentious and are open to debate and dispute. No longer can students graduating from qualifying courses expect to go into local authority social work posts. Students are now just as likely to find employment in the independent sector in either voluntary organisations, private agencies or within one of the multitude of multi-agency teams in youth justice, learning disabilities or mental health. This is also likely to lead to social workers being employed in multiprofessional as opposed to uni-professional teams. This trend is likely to continue as is the governmental demands for managerial control over practice to be reflected in star ratings informed by standards and performance indicators. These drivers could be seen to weaken the position of social work as a profession and challenge social work research as a key discipline that seeks to inform practice.

The future is notoriously difficult to predict and the certainties of one age, for example steam engines, become outmoded in the next. One thing though is for certain:

> Doing more of the same won't work. Increasing demand, greater complexity and rising expectations mean that the current situation is not sustainable. (Roe, 2006: 8)

The above quote from the *Report of the 21st Century Social Work Review: Changing Lives* aptly illustrates the point that social work cannot stand still; it must be responsive to the context in which it operates and relevant to those whom it wishes to serve.

Social work is experiencing a period of significant change. For the first time in the UK social work is an all-graduate workforce with a protected title and a requirement to register with the requisite care council. To qualify as a social worker students will have to 'demonstrate professional competence in social work practice' (DH, 2000b: key role 6) and it is difficult to see how they will be able to achieve this if they are unable to analyse, evaluate and use current research evidence of effective social work practice. Registration will not be a once and for all activity as those registered will be expected to demonstrate they have undertaken continuous professional development activities to be able to re-register whilst there is also a post-qualifying structure for social workers at degree and postgraduate levels.

In recognition of the importance of research-informed practice the English Government has created SCIE whilst the Scottish Executive has also developed reSearch Web (www.researchweb.org.uk) to promote research getting into practice. Local authorities and voluntary agencies have been signing up to national organisations like RiP, RiPfA and MRC in an acknowledgement that research is desirable and can impact upon better practice. Research has never had such a prominent place in social work. Managers and government are keen to promote evidence-based practice and the evaluation of individual practitioner performance. Practitioner research is well accepted and seen as important to both agencies and practitioners. The future remains imperfect, our only guarantee is that change is here to stay. Social work, if it is to survive, has to work within and outside the context in which it is located. It has to explore the possibilities of the context for creative practice and at the same time to identify ways in which to move towards other futures which highlight the moral nature of the social work task, acknowledge and embrace the growing importance of service user involvement and highlight anti-oppressive practice and inter-agency working. No professional group, and this is as true for social work as it is for lawyers, doctors or ministers, has an unequivocal right to survival. If social work is to survive it must engage with contemporary debates and seek to shape, as much as being shaped by, external events. In seeking to engage in this debate this book has highlighted the importance of the contribution that social work research can make to the future of social work practice and vice versa. The book has argued for the research-minded practitioner and the research literate workforce that is able to capitalise on the insights from research, but also able to distinguish between knowledge that is valid and credible and knowledge that is not. Social work research can potentially become a key tool for social workers in seeking to attest to the profession's professionalism and support it in its attempts to achieve a transformative practice dealing both with the issues of social control and care.

Knowledge production and research-informed practices are important, but social work research also has to strengthen its standing amongst the other disciplines within the university. One of the distinguishing features of social work research is its applied nature, but that does not mean it should eschew basic research. Basic research helps to provide the conceptual blocks for understanding practice and developing ways in which those blocks can be used to transform practice developing new ways of working that empower service users and social workers. For social work research there is an inherent requirement to be practice oriented but also for it to be effective it must be recognised within the academy and wider society as a legitimate knowledge-producing discipline. This represents the dual challenges for social work research – for it to prosper it must be able to contribute to both arenas, not one or the other, but both.

Never before has it been so important for social work research to emphasise what Butler (2003: 27) has described as a 'value based, politically aware and engaged form of endeavour'. Such a stance will inevitably push social work research into conflict with other parts of society and the academic world. However, that is to be expected if social work research is to demonstrate a respect for persons, promote user self-determination, work for the interests of service users and promote social justice. For social workers to be able to act effectively in these domains they need to be able to work with all the tools at their disposal. It is one of the key arguments of this book that good social workers make use of social work research. It is also a key message from this book that social work research is a key tool in the social worker's toolbox that is all too often underused or misused.

Appendix: Abbreviations used in the Book

BASW	British Association of Social Workers. Professional membership organisation for qualified social workers.
CAFCASS	Children and Family Court Advice and Support Services. Looks after the interests of children involved in family proceedings including adoption and divorce.
CCETSW	Central Council for Education and Training in Social Work. Forerunner of the General Social Care Council.
CSCI	Commission for Social Care Inspection. Single independent inspection body for social care in England.
DH	Department of Health. Responsible for adult social care services.
Dip SW	Diploma in Social Work. The professional qualification for social workers prior to the introduction of the new award in 2003.
EBP	Evidence based practice. A very important and contested concept for the promotion of research-informing practice.
ECDL	European Computer Driving Licence. A recognised standard of computer literacy that newly qualifying social workers have had to achieve.
ESRC	Economic and Social Research Council. A major provider of research grants.
GSCC	General Social Care Council. The registration and regulatory body for social workers and social care.
HEFC	Higher Education Funding Council. Provides funding for higher education.
IASSW	International Association of Schools of Social Work. An international academic organisation for promoting social work education.
IFSW	International Federation of Social Workers. International campaigning organisation for social workers.
JAR	Joint Area Reviews. CSCI along with nine other inspectorates' inspection of an area's Children's Services including those services provided by social workers.

MRC	Making Research Count. A national research development initiative based on a hub and spoke principle between universities, local authorities and voluntary agencies.
PAF	Performance Assessment Framework. A programme of key performance measures for measuring social services' performance.
PQ	Post-Qualifying. Refers to the social work qualification framework beyond the qualifying degree.
QAA	Quality Assurance Agency. There is a specific Quality Assurance Agency for Higher Education whose role is to ensure higher education courses meet subject benchmarks and are of a suitable standard.
RAE	Research Assessment Exercise. Regular research selectivity exercise used to allocate university research funds.
RCT	Random Controlled Trial. Specialist and some would say 'gold standard' research method.
RGF	Research Governance Framework. DH initiative to ensure local authority social care research is ethical and accountable.
RiP	Research in Practice. A national organisation promoting the utilisation of research for practice with a very informative website – www.rip.org.uk.
RiPfA	Research in Practice for Adults. Sister organisation of RiP for promoting research utilisation in adult social care.
SCIE	Social Care Institute for Excellence. Collects and synthesises up-to-date knowledge about what works in social care and then makes this knowledge available and accessible.
SIESWE	Scottish Institute of Excellence in Social Work Education. Established by the nine universities providing social work qualifying training in Scotland to identify and disseminate best practice.
TOPSS	Training Organisation for Personal Social Services. Has been replaced by Skills for Care, a national employer led sector skills council.
UPIAS	Union of Physically Impaired Against Segregation. A campaigning organisation for people with disabilities.
YOT	Youth Offending Team. Inter-agency teams constructed to manage and reduce youth offending consisting of representatives from social work, probation, police, education and health services personnel.

References

Ackroyd, S. and Hughes, J. A. (1981) *Data Collection in Context.* London: Longman.

Adams, R. (1998) *Quality Social Work.* Basingstoke: Macmillan.

Adams, R. (2002) *Social Policy for Social Work.* Basingstoke: Palgrave.

Ahmad, B. (1990) *Black Perspectives in Social Work.* Birmingham: Venture Press.

Alcock, P. and Williams, B. (2000) *Social Policy and Administration and Social Work: Subject Benchmark Statements.* Gloucester: Quality Assurance Agency for Higher Education.

Antman, E., Lau, J., Kupeltruck, B., Mostellet, F. and Chalmers, I. (1992) 'A comparison of the results of meta-analyses of random controlled trials and recommendations of clinical experts', *Journal of the American Medical Association,* 268(4): 240–8.

Argyris, C. and Schon, D. A. (1978) *Organisational Learning: A Theory to Action Perspective.* Reading: Addison Wesley.

Arnstein, S. (1971) 'A ladder of citizen participation', *Journal of the Royal Planning Institute,* 35(4): 216–24.

Atkin, K. and Rollings, J. (1993) *Community Care in a Multi-Racial Britain.* London: HMSO.

Bailey, R. and Brake, M. (1975) *Radical Social Work.* London: Arnold.

Baldwin, M. (2004) 'Optimism and the art of the possible', in N. Gould and M. Baldwin (eds), *Social Work: Critical Reflection and the Learning Organization.* Aldershot: Ashgate. pp. 161–76.

Balloch, S. and Taylor, M. (2001) *Partnership Working: Policy and Practice.* Bristol: The Policy Press.

Banks, S. (1995) *Ethics and Values in Social Work.* Basingstoke: Macmillan.

Barber, B. (1976) 'The ethics of experimentation with human subjects', *Scientific American,* 234(2): 25–31.

Barclay, P. (1982) *Social Workers: Their Roles and Tasks.* London: NISW/Bedford Square Press.

Barn, R. (1993) *Black Children in the Public Care System.* London: Batsford.

Barn, R. (1994) 'Black children in the public care system', in B. Humphries and C. Truman (eds), *Re-Thinking Social Research: Anti-Discrimination Approaches in Research Methodology*. Aldershot: Avebury. pp. 37–58.

Barnes, C. and Mercer, G. (1997) 'Breaking the mould: an introduction to disability research', in C. Barnes and G. Mercer (eds), *Doing Disability Research*. Leeds: The Disability Press.

Barnes, K. (2002) *Focus on the Future: Key Messages From Focus Groups on the Future of Social Work Training*. London: Department of Health.

Barr, H. (2003) 'Unpacking interprofessional education', in A. Leathard (ed.), *Interprofessional Collaboration: From Policy to Practice in Health and Social Care*. Hove: Brunner-Routledge.

Barr, H., Freeth, D., Hammick, M., Koppell, I. and Reeves, S. (1999) *Evaluating Interprofessional Education: A United Kingdom Review for Health and Social Care*. London: BERA/CAIPE.

Barratt, M. and Hodson, R. (2006) *Firm Foundations: A Practical Guide to Organisational Support for the Use of Research Evidence*. Dartington: Research in Practice.

BASW (2002) *The Code of Ethics for Social Work*. www.basw.co.uk/articleld=2fpage=1, accessed 10 Sept. 2005.

Bateson, G. (1972) *Steps to an Ecology of Mind*. New York: Ballantine.

Bebington, A. and Miles, J. (1989) 'The background of children who enter local authority care', *British Journal of Social Work*, 19(5): 349–68.

Becker, S. and Bryman, A. (2004) *Understanding Research for Social Policy and Practice: Themes, Methods and Approaches*. Bristol: The Policy Press.

Begum, N., Hill, M. and Stevens, A. (eds) (1994) 'Reflections: views of black disabled people on their lives and community care'. Central Council for Education and Training of Social Workers, Paper 32.3, London.

Beresford, P. (1994) *Changing the Culture: Involving Service Users in Social Work Education*. London: CCETSW.

Beresford, P. and Croft, S. (1993) *Citizen Involvement*. London: Macmillan.

Berger, P. and Luckman, T. (1979) *The Social Construction of Reality*. London: Peregrine.

Biestek, F. (1961) *The Casework Relationship*. London: George Allen and Unwin.

Booth, T. (1988) *Developing Policy Research*. London: Avebury.

Boushel, M. (1996) 'Vulnerable multi-racial families and early years services: concerns, challenges and opportunities', *Children and Society*, 10: 305–16.

Boushel, M. (2000) 'What kind of people are we? "Race", anti-racism and social welfare research' *British Journal of Social Work*, 30(1): 71–89.

Brandt, A. M. (1978) *Racism, Research and the Tuskegee Syphilis Study (Report No. 8)*. New York: Hastings Center.

Bullock, R., Gooch, D., Little, M. and Mount, K. (1998) *Research in Practice: Experiments in Development and Information Design*. Aldershot: Ashgate.

Butler, I. (2002) 'A code of ethics for social work and social care research', *British Journal of Social Work*, 32(2): 239–48.
Butler, I. (2003) 'Doing good research and doing it well', *Social Work Education*, 33(1): 19–30.
Butler, I., Scanlan, L., Robinson, M., Douglas, G. and Murch, M. (2002) 'Children's involvement in their parents' divorce: implications for practice', *Children and Society*, 16(1): 89–102.
Butrym, Z. (1976) *The Nature of Social Work*. London: Macmillan.
Butt, J. and Mizra, K. (1996) *Social Care and Black Communities: A Review of Recent Research Studies*. London: HMSO.
Butt, J. and O'Neill, A. (2005) *'Let's Move On': Black and Minority Ethnic Older People's Views on Research Findings*. York: Joseph Rowntree Foundation.
Cabinet Office (1990) *Modernising Government (Cm4310)*. London: Stationery Office.
Cabot, R. C. (1931) 'Treatment in social casework and the need for tests of its success and failure'. Proceedings of the National Conference of Social Work. USA.
Campbell, D. T. and Fiske, D. W. (1959) 'Convergent and discriminant validation by the multitrait-multimethod matrix', *Psychological Bulletin*, 56(1): 81–105.
Carr, S. (2004) *Has Service User Participation Made a Difference to Social Care Services?* London: SCIE and the Policy Press.
CCETSW (1976) *Values in Social Work (Paper 13)*. London: Central Council for Education and Training in Social Work.
CCETSW (1989) *Rules and Requirements for the Diploma in Social Work*. London: Central Council for the Education and Training in Social Work.
CCETSW (1991a) *Rules and Requirements for the Diploma in Social Work (Paper 30)*. London: Central Council for the Education and Training in Social Work.
CCETSW (1991b) *Setting the Context for Change: Anti-Racist Social Work Education*. London: Central Council for the Education and Training in Social Work.
CCETSW (1995) *Rules and Requirements for the Diploma in Social Work (Revised)*. London: Central Council for the Education and Training in Social Work.
Chapman, T. and Hough, M. (eds) (1998) *Evidence-based Practice: A Guide to Effective Practice*. London: Home Office.
Charles, M., Rashid, S. and Thoburn, J. (1992) 'The placement of black children with permanent new families', *Adoption and Fostering*, 16(1): 13–19.
Children and Young People's Unit (2001) *Learning to Listen: Core Principles for the Involvement of Children and Young People*. London: CYPU.
Christensen, P. and Prout, A. (2002) 'Working with ethical symmetry in social research with children', *Childhood*, 9(4): 477–97.
Citizens as Trainers, Young Independent People Presenting Educational Entertainment, Rimmer, A. and Harwood, K. (2004) 'Citizen participation in the education and training of social workers', *Social Work Education*, 22(3): 309–23.

Clark, A. (2004) 'The mosaic approach and research with young children', in S. Fraser, V., Lewis, S. Ding, M. Kellett, and C. Robinson (eds), *The Reality of Research with Children and Young People.* London: Sage in association with the Open University Press. pp. 142–56.

Clifford, D. (1994) 'Critical life histories: key anti-oppressive research methods and processes', in B. Humphries and C. Truman, (eds), *Re-Thinking Social Research: Anti-Discriminatory Approaches in Research Methodology.* Aldershot: Avebury. pp. 102–22.

Coghlan, D. and Brannick, T. (2001) *Doing Action Research: In Your Own Organization.* London: Sage.

Corrigan, P. and Leonard, P. (1978) *Critical Texts in Social Work and the Welfare State.* London: Macmillan.

Croft, S. and Beresford, P. (1989) 'User involvement, citizenship and social policy', *Critical Social Policy,* 26(1): 5–18.

Cronin, A. (2001) 'Focus groups', N. Gilbert (ed.), *Researching Social Life* (2nd edn), London: Sage.

Dalrymple, J. and Burke, B. (1995) *Anti-oppressive Practice.* Buckingham: Open University Press.

Davey, B., Levin, E., Ilife, S. and Khariccah, K. (2005) 'Integrating health and social care: implications for joint working and community care outcomes for older people', *Journal of Interprofessional Care,* 19(1): 22–34.

Davidson, J. O. C. and Layder, D. (1994) *Methods, Sex and Madness.* London: Routledge.

Day, L. (1993) 'Women and oppression: race, class and gender', in M. Langan and L. Day (eds), *Women, Oppression and Social Work.* London: Routledge. pp. 12–31.

D'Cruz, H. and Jones, M. (2004) *Social Work Research: Ethical and Political Contexts.* London: Sage.

Deetz, S. (1993) *Communication 2000: The Discipline, the Challenges, the Research, the Social Contribution.* New Jersey: Rutgers University.

Delanty, G. and Strydom, P. (2003) 'Introduction: what is the philosophy of social science?' in G. Delanty and P. Strydom (eds), *Philosophies of Social Science: The Classic and Contemporary Readings.* Maidenhead: Open University.

Denzin, N. K. (1989) *Interpretive Interactionism.* London: Sage.

Department of Health (1998) *Modernising Social Services: Promoting Independence, Improving Protection and Raising Standards (CM4169).* London: The Stationery Office.

Department of Health (1999a) *Framework for the Assessment of Children in Need and their Families.* London: HMSO.

Department of Health (1999b) *National Service Framework for Mental Health: Modern Standards and Modern Service Models.* London: Department of Health.

Department of Health (2000a) *A Quality Strategy for Social Care.* London: Department of Health.

Department of Health (2000b) *The NHS Plan: A Plan for Investment, A Plan for Reform.* London: Department of Health.

Department of Health (2001a) *Fair Access to Care Services.* London: Stationery Office.

Department of Health (2001b) *Research Governance Framework for Health and Social Care.* London: Department of Health.

Department of Health (2001c) *National Service Framework for Older People.* London: Department of Health.

Department of Health (2001d) *Valuing People: A New Strategy for Learning Disability for the 21st Century.* London: Department of Health.

Department of Health (2001e) *Mental Health National Service Framework: Full Report by the Workforce Action Team on Workforce Planning, Education and Training.* London: HMSO.

Department of Health (2001f) *Working Together Learning Together: A Framework for Lifelong Working for the NHS.* London: HMSO.

Department of Health (2002a) *Requirements for Social Work Training.* London: Department of Health.

Department of Health (2002b) *Listening, Hearing and Responding.* London: Stationery Office.

Department of Health (2004) *The Research Governance Framework for Health and Social Care.* London: Department of Health.

Department of Health and Department for Education and Skills (2004) *National Service Framework for Children, Young People and Maternity Services: Executive Summary.* London: Department of Health.

DETR (1998) *Modern Local Government – In Touch with the People (Cm4014).* London: Stationery Office.

DoH/RiP/MRC (2002) *Quality Protects Research Briefing: Placement Stability.* London: DoH/RiP/MRC.

Dominelli, L. (1993) *Social Work: Mirror of Society or its Conscience?* Sheffield: Department of Sociological Studies.

Dominelli, L. (1997) *Anti-Racist Social Work.* Basingstoke: Macmillan.

Dominelli, L. (1998) 'Anti-oppressive practice in context', in R. Adams, L. Dominelli and M. Payne (eds), *Social Work: Themes, Issues and Critical Debates.* Basingstoke: Macmillan Press. pp. 3–22.

Dominelli, L. (2005) 'Social work research: contested knowledge for practice', in R. Adams, L. Dominelli and M. Payne (eds), *Social Work Futures: Crossing Boundaries, Transforming Practice.* Basingstoke: Palgrave Macmillan. pp. 223–36.

Dominelli, L. and McLeod, E. (1989) *Feminist Social Work.* Basingstoke: Macmillan.

Downie, R. S. and Telfer, E. (1969) *Respect for Persons.* London: Allen and Unwin.

Easterby-Smith, M. and Thorpe, R. (1996) *Research Traditions in Management Learning.* London: Sage.

Easterby-Smith, M., Thorpe, R. and Lowe, A. (1991) *Management Research: An Introduction.* London: Sage.

England, H. (ed.) (1986) *Social Work as Art.* London: Allen and Unwin.

ESRC (2006) *Annual Report And Accounts 2004–5.* Swindon: ESRC.

Evans, C. and Carmichael, A. (2002) *User's Best Value: A Guide to Good Practice in User Involvement in Best Value Reviews.* York: Joseph Rowntree Foundation, and Wiltshire: Swindon Service Users' Network and the University of Bath.

Everitt, A., Hardiker, P., Litlewood, J. and Mullender, A. (1992) *Applied Research for Better Practice.* Basingstoke: Macmillan.

Farmer, E. and Owen, M. (1995) *Child Protection Practice: Private Risks, Public Remedies, A Study of Decision-Making, Intervention and Outcomes in Child Protection Work.* London: HMSO.

Finklestein, V. (1985) Unpublished paper at the World Health Organisation, Netherlands.

Folkard, S., Smith, D. E. and Smith, D. D. (1976) *IMPACT Intensive Matched Probation and After-Care Treatment. Vol. 11: The Results of the Experiment.* London: HMSO.

Foster, P. (1990) *Policy and Practice in Multicultural and Anti-Racist Education.* London: Routledge.

Foster, P. (1991) 'Cases not proven: a reply to Cecile Wright', *British Educational Research Journal,* 17(2): 335–48.

Foster, P. (1992) 'What are Connolly's rules? A reply to Paul Connolly', *British Educational Research Journal,* 18(2): 149–54.

France, A. (2004) 'Young people', in S. Fraser, J. Lewis, S. Ding, M. Kellett and C. Robinson (eds), *Doing Research with Young People.* London: Sage in association with the Open University. pp. 97–112.

Frost, N. (2005) *Professionalism, Partnership and Joined -Up Working.* Dartington: Research in Practice.

Furedi, F. (2002) 'Don't rock the research boat', *Times Higher Education Supplement,* 11 January: 20.

Garfinkel, H. (1967) *Studies in Ethnomethodology.* Englewood Cliffs, NJ: Prentice Hall.

Garrett, P. M. (2000) 'Responding to Irish "invisibility": anti-discriminatory social work practice and the placement of Irish children in Britain', *Adoption and Fostering,* 24(1): 23–34.

Garrett, P. M. (2003) *Remaking Social Work with Children and Families: A Critical Discussion on the 'Modernisation' of Social Care.* London: Routledge.

Gibbons, M., Limoges, C., Nowotny, H., Schwartzman, S., Scott, P. and Trow, M. (1994) *The New Production of Knowledge: The Dynamics of Science and Research in Contemporary Societies.* London: Sage.

Gibbons, J., Conroy, S. and Bell, C. (1995) *Operating Child Protection Policies in English Local Authorities*. London: HMSO.

Giddens, A. (1977) *Studies in Social and Political Theory*. London: Hutchinson.

Giddens, A. (2001) *Sociology*. Cambridge: Polity Press.

Gilbert, N. (2001) 'Research theory and method', in N. Gilbert (ed.), *Researching Social Life*. London: Sage. pp. 14–27.

Gillon, R. (1994) 'Medical ethics: four principles plus attention to scope', *British Medical Journal*, 309: 184–8.

Glasby, J. and Lester, H. (2004) 'Cases for change in mental health: partnership working in mental health services', *Journal of Interprofessional Care*, 18(1): 7–16.

Glendinning, C., Powell, M. and Rummery, K. (eds) (2002) *Partnerships, New Labour and the Governance of Welfare*. Bristol: The Policy Press.

GNCs/CCETSW (1982) *Co-operation in Training Part 1. Qualifying Training*. London: General Nursing Councils for England and Wales, Scotland and Northern Ireland and the Central Council for Education and Training in Social Work.

Goffman, E. (1961) *Asylums*. Garden City, NJ: Doubleday Anchor Books.

Gopineth, C. and Hoffman, C. (1995) 'The relevance of strategy research: practitioner and academic views', *Journal of Management Studies*, 32(5): 575–94.

Gould, N. (2004) 'Introduction: the learning organization and reflective practice – the emergence of a concept', in N. Gould and M. Baldwin (eds), *Social Work. Critical reflection and the Learning Organization*. Aldershot: Ashgate, pp. 1–9.

GSCC (2002) *Codes of Practice for Employers of Social Care Workers and for Social Care Workers*. London: GSCC.

GSCC (2005) *Post Qualifying Framework for Social Work Education and Training*. London: General Social Care Council.

Guena, A., Hidayat, O. and Martin, B. (1999) *Resource Allocation and Research Performance: The Assessment of Research*. Brighton: HEFCE.

Gutek, G. A. (1978) 'Strategies for studying client satisfaction', *Journal of Social Issues*, 34(4): 44–55.

Hammersley, M. (1995) *The Politics of Social Research*. London: Sage.

Hanley, B. (2005) *Research as Empowerment? Report of a Series of Seminars Organised by the Toronto Group*. York: Joseph Rowntree Foundation.

Hanley, B., Bradburn, J., Barnes, M., Evans, C., Goodare, H., Kelson, M., Kent, A., Oliver, S., Thomas, S. and Wallcraft, J. (2004) *Involving the Public in NHS, Public Health and Social Care Research: Briefing Notes for Researchers*. Eastleigh: Involve.

Harding, T. and Beresford, P. (1996) *The Standards We Expect: What Service Users and Carers want from Social Service Workers*. London: National Institute for Social Work.

Harvey, L. (1990) *Critical Social Research.* London: Unwin Hyman.

Hasler, F. (2003) *Users at the Heart: User Participation in the Governance and Operation of Social Regulatory Bodies.* London: SCIE.

Healy, K. (2005) *Social Work Theories in Context: Creating Frameworks for Practice.* Basingstoke: Palgrave Macmillan.

Heath, A., Colton, M. and Aldgate, J. (1994) 'Failure to escape: A longitudinal study of foster children's educational attainment', *British Journal of Social Work*, 24(3): 241–60.

Heron, J. (1996) *Co-operative Inquiry: Research into the Human Condition.* London: Sage.

Hewson, C., Yule, P., Laurent, D. and Vogel, C. (2003) *Internet Research Methods: A Practical Guide for the Social and Behavioural Sciences.* London: Sage.

Higham, P. (2001) *Integrated Practice: Using Narratives to Develop Theories in Social Care.* Birmingham: Venture Press.

Homan, R. (1991) *The Ethics of Social Research.* Harlow: Longmans.

Horner, N. (2003) *What is Social Work? Context and Perspectives.* Poole: Learning Matters.

Hudson, B. (2002) 'Interprofessionality in health and social care: the Achilles' heel of partnership?', *Journal of Interprofessional Care*, 16(1): 7–17.

Hughes, J. (1990) *The Philosophy of Social Research.* London: Longman.

Hugman, R. (1991) *Power in Caring Professions.* Basingstoke: Macmillan.

Hugman, R. (2003) 'Going around in circles? Identifying interprofessional dynamics in Australian health and social welfare', in A. Leathard (ed.), *Interprofessional Collaboration: From Policy to Practice in Health and Social Care.* Hove: Brunner-Routledge, pp. 56–68.

Hugman, R. and Smith, D. (1995) 'Ethical issues in social work: an overview', in R. Hugman, and D. Smith (eds), *Ethical Issues in Social Work.* London: Routledge. pp. 1–15.

Humphreys, L. (1975) *Tearoom Trade: Impersonal Sex in Public Places.* Chicago: Aldine.

Humphries, B. (1996) 'Contradictions in the culture of empowerment', in B. Humphries (ed.), *Critical Perspectives on Empowerment.* Birmingham: Venture Press. pp. 1–16.

Humphries, B. (2004) 'An unacceptable role for social work: implementing immigration policy', *British Journal of Social Work*, 34(1): 93–107.

Humphries, B. (2005) 'From margin to centre: shifting the emphasis of social work research', in R. Adams, L. Dominelli, and M. Payne (eds), *Social Work Futures: Crossing Boundaries, Transforming Practice.* Basingstoke: Palgrave Macmillan. pp. 279–92.

Humphries, B. and Martin, M. (2000) 'Disrupting ethics in social research', in B. Humphries (ed.), *Research in Social Care and Social Welfare: Issues and Debates for Practice.* London: Jessica Kingsley.

Hunt, P. (1981) 'Settling accounts with the parasite people', *Disability Challenge*, 2(1): 37–50.

Husband, C. (1995) 'The morally active practitioner and the ethics of anti-racist social work', in R. Hugman and D. Smith (eds), *Ethical Issues in Social Work*. London: Routledge. pp. 84–103.

IASSW and IFSW (2004) Global Standards for Social Work Education, at www.iassw.soton.ac.uk/en/GlobalQualifyingStandards/GlobalStandards/pdf, accessed 28 Aug. 2005.

Jackson, S. (1987) *The Education of Children in Care*. Bristol Papers in Applied Social Studies no.1: University of Bristol.

Jackson, S. (1994) 'Educating children in residential and foster care', *Oxford Review of Education*, 20(2): 267–79.

Jay Committee (1979) *Report of the Committee of Enquiry into Mental Handicap Nursing and Care*. London: HMSO.

Johns, R. (2003) *Using the Law in Social Work*. Exeter: Learning Matters.

Jones, R. (1995) 'Co-opting carers and service users', *ADSS News*, April: 18–19.

Kadushin, A. (1972) *The Social Work Interview*. London: Columbia University Press.

Kimmel, A. J. (1988) *Ethics and Values in Applied Social Research*. Newbury Park: Sage.

Kirby, P. (2004) *A Guide to Actively Involving Young People in Research: For Researchers, Research Commissioners and Managers*. Eastleigh: Involve.

Kuhn, T. (1970) *The Structure of Scientific Revolutions* (2nd edn). London: University of Chicago Press.

Langston, A., Abbott, L., Lewis, V. and Kellett, M. (2004) 'Early childhood', in S. Fraser, J. Lewis, S. Ding, M. Kellett and C. Robinson (eds), *Doing Research with Children and Young People*. London: Sage in association with Open University. pp. 147–60.

Law, S. and Janzon, K. (2005) 'Reaching for the stars: the performance assessment framework for social services', in D. Taylor and S. Bulloch (eds), *The Politics of Evaluation: Participation and Policy Implementation*. Bristol: The Policy Press. pp. 57–73.

Leathard, A. (1994) 'Inter-professional developments in Britain. An overview', in A. Leathard, *Going Inter-professional. Working Together for Health and Welfare*. London: Routledge.

Leathard, A. (2003) 'Introduction' in A. Leathard (ed.), *Interprofessional Collaboration: From Policy to Practice in Health and Social Care*. Hove: Brunner-Routledge.

Lewis, J. (2002) 'Research and development in social care: governance and good practice', *Research Policy and Planning*, 20(1): 3–10.

Lewis, J. (2003) 'Design issues', in J. Ritchie and J. Lewis (eds), *Qualitative Research Practice: A Guide for Social Science Students and Researchers*. London: Sage. pp. 47–76.

Lincoln, Y. S. and Denzin, N. (1998) 'The fifth movement', in N. Denzin and Y. S. Lincoln (eds), *The Landscape of Qualitative Research.* London: Sage, 407–30.

Lockey, R., Sitzia, J., Gillingham, T., Millyard, T., Miller, C., Ahmed, S., Beales, A., Bennett, C., Parfoot, S., Sigrist, J. and Worthing and Southlands Hospitals NHS Trust (2004) *Training for Service Users Involvement in Health and Social Care Research: A Study of Training Provision and Participants Experiences.* Eastleigh: Involve.

Lorde, A. (1984) *Sister Outsider.* New York: Crossing Press.

Lymbery, M. (1998) 'Social work and general practice: dilemmas and solutions', *Journal of Interprofessional Care,* 12(2): 199–208.

MacDonald, G. (1996) 'Ice therapy: why we need randomised control trials', in P. Alderson, S. Brill and I. Chalmers (eds), *What Works: Effective Social Interventions in Child Welfare.* Barkingside: Barnados.

MacDonald, G. (1999) 'Social work and its evaluation: a methodological dilemma', in F. Williams, J. Popay and A. Oakley (eds), *Welfare Research: A Critical Review.* London: UCL Press. pp. 89–103.

MacDonald, G. (2003) *Using Systematic Reviews to Improve Social Care: Report No.4.* London: SCIE.

MacDonald, G. and Roberts, H. (1995) *What Works in the Early Years.* Ilford: Barnados.

MacDonald, G. and Winkley, A. (1999) *What Works in Child Protection.* Ilford: Barnados.

MacDonald, G., Sheldon, B. and Gillespie, J. (1992) 'Contemporary studies of the effectiveness of social work', *British Journal of Social Work,* 22(6): 615–42.

MacIntyre, A. (1985) *After Virtue,* London: Duckworth.

Macey, M. and Moxon, E. (1996) 'An examination of anti-racist and anti-oppressive theory and practice in social work education', *British Journal of Social Work,* 26(3): 297–314.

Mackie, J. L. (1977) *Ethics: Inventing Right and Wrong.* Harmondsworth: Penguin.

Marsh, P and Fisher, M. in collaboration with Mathers, N. and Fish, S. (2005) *Developing the Evidence Base for Social Work and Social Care Practice.* London: SCIE.

May, T. (2001) *Social Research: Issues, Methods and Process* (3rd edn). Buckingham: Open University Press.

May, T. and Williams, M. (2001) 'Social surveys: design to analysis', in T. May (ed.), *Social Research: Issues methods and process.* Buckingham: Open University Press. pp. 88–119.

Mayer, J. E. and Timms, N. (1970) *The Client Speaks.* London: Routledge and Kegan Paul.

McDonald, P. and Coleman, M. (1999) 'Deconstructing hierarchies of oppression and adopting a "multi model" approach to anti-oppressive practice,' *Social Work Education,* 18(1): 19–33.

McGuire, J. (1995) *What Works: Reducing Offending.* Chichester: Wiley.

McLaughlin, H., Brown, D. and Young, A. (2004) 'Consultation, community and empowerment: lessons from the deaf community', *Journal of Social Work,* 4(2): 153–65.

McLaughlin, H. (2005) 'Young service users as co-researchers', *Qualitative Social Work,* 4(1): 21–8.

Merrington, S. and Stanley, S. (2000) 'Doubts about the "what works" initiative' *Probation Journal,* 47(6): 272–5.

Merrington, S. and Stanley, S. (2004) '"What works?": revisiting the evidence in England and Wales', *Probation Journal,* 51(1): 7–20.

Miller, E. J. and Gwynne, G. V. (1972) *A Life Apart.* London: Tavistock.

Mills, D., Jepson, P., Coxon, T., Easterby-Smith, M., Hawkins, P. and Spencer, J. *Demographic Review of the UK Social Sciences.* Swindon: ESRC. www.esrc.ac.uk/ESRCIfoCentre/Images/Demographic_Review_tem6–13872.pdf accessed 06/04/2006

Moffet, J. (1968) *Concepts of Casework Treatment.* London: Routledge and Kegan Paul.

Molyneux, J. and Irvine, J. (2004) 'Service users and carers involvement in social work training: the long and winding road', *Social Work Education,* 23(3): 293–308.

Muluccio, A. (1998) 'Assessing child welfare outcomes: the American perspective', *Children and Society,* 12(2): 161–8.

Newman, J. (2000) 'Beyond the new public management? Modernizing public services', in J. Clarke, S. Gerwitz and E. McLaughlin (eds), *New Managerialism New Welfare.* London: Open University in association with Sage.

Newman, T., Moseley, A., Tierney, S. and Ellis, A. (2005) *Evidence-based Social Work: A Guide for the Perplexed.* Lyme Regis: Russell House Publishing.

Nutley, S. and Webb, J. (2000) 'Evidence and the policy process', in H. N. Davies, and P. Smith (eds), *What Works? Evidence-based Policy and Practice in Public Services.* Bristol: The Policy Press. pp. 13–41.

Oakley, A. (1981) 'Interviewing women: a contradiction in terms', in H. Roberts (eds.), *Doing Feminist Research.* London: Routledge and Kegan Paul. pp. 30–61.

Oates, R. K. and Bross, D. C. (1995) 'What have we learned from treating physical abuse?', *Child Abuse and Neglect,* 19, 463–73.

Oliver, M. (1990) *The Politics of Disablement.* London: Macmillan.

Oliver, M. (1993) 'Re-defining disability: a challenge to research', in J. Swann, V. Finkelstein, S. French and M. Oliver (eds), *Disabling Barriers: Enabling Research.* London: Sage, pp. 61–7.

Oliver, P. (2003) *The Student's Guide to Research Ethics.* Maidenhead: Open University Press.

Orme, J. (2003) '"It's Feminist because I say so!" Feminism, social work and critical practice in the UK', *Qualitative Social Work,* 2(2): 218–26.

Ovretveit, J. (1993) *Co-ordinating Community Care. Multidisciplinary teams and Care Management.* Buckingham: Open University Press.

Ovretveit, J. (1997) 'How patient power and client participation affects relations between professions', in J. Ovretveit, P. Mathias and T. Thompson (eds), *Interprofessional Working for Health and Social Care,* London: Macmillan. pp. 79–102.

Paley, J. (2000) 'Paradigms and presuppositions: the difference between qualitative and quantitative research', *Scholarly Inquiry for Nursing Practice,* 14(2): 143–55.

Parker, R. (1966) *Decisions in Child Care: A Study of Prediction in Fostering.* London: Allen and Unwin.

Parker, R. (1998) 'Evidence, judgment and values', in J. Tunnard (ed.), *Commissioning and Managing External Research: A Guide for Child Care Agencies.* Dartington: Research in Practice. pp. 5–12.

Parker, R., Ward, H., Jackson, S., Aldgate, J. and Wedge, P. (1991) *Looking After Children: Assessing Outcomes in Childcare.* London: HMSO.

Pawson, R., Boaz, A., Grayson, L., Long, A. and Barnes, C. (2003) *Types and Quality of Knowledge in Social Care.* London: SCIE.

Payne, G. and Payne, J. (2004) *Key Concepts in Social Research.* London: Sage.

Payne, M. (1997) *Modern Social Work Theory* (2nd edn). Basingstoke: Macmillan.

Payne, M. (2000) *Anti-Bureaucratic Social Work.* Birmingham: Venture Press.

Payne, M. (2005) *Modern Social Work Theory* (3rd edn). Basingstoke: Palgrave Macmillan.

Payne, M and Shardlow, S. M. (eds) (2002) *Social Work in the British Isles.* London: Jessica Kingsley.

Pedler, M., Burgoyne, J. and Boydell, T. (1991) *The Learning Company.* London: McGraw-Hill.

Pew Health Professions Commission (1995) *Critical Challenges: Revitalising the Health Professions for the Twenty-First Century.* San Francisco, CA: Pew Health Professions Commission.

Plant, R. (1970) *Social and Moral Theory in Casework.* London: Routledge and Kegan Paul.

Platt, D. (2002) *Modern Social Services: A Commitment to Reform.* London: Stationery Office.

Popper, K. (1980) *The Logic of Scientific Discovery.* London: Hutchinson.

Preston-Shoot, M. (2002) 'Why social workers don't read', *Care and Health Guide,* 11 March: 11–12.

Pugh, R. (2003) 'Considering the countryside: is there a case for rural social work?', *British Journal of Social Work,* 33(1): 67–86.

Punch, M. (1998) 'Politics of ethics in qualitative research', in N. Denzin and Y. S. Lincoln (eds), *The Landscape of Qualitative Research.* London: Sage. pp. 156–84.

Qvortrup, J., Bardy, M., Sgritta, G. and Wintersberger, H. (eds) (1994) *Childhood Matters.* Vienna: European Centre.

Ragg, N. (1977) *People Not Cases.* London: Routledge and Kegan Paul.

Ratcliffe, P. (2004) *'Race', Ethnicity and Difference: Imagining the Inclusive Society.* Maidenhead: Open University Press.

Raynor, P. (2003) 'Evidence-based practice and its critics', *Probation Journal,* 50(3): 334–45.

Reason, P. and Rowan, J. (eds) (1981) *Human Inquiry: A Sourcebook of New Paradigm Research.* London: Wiley.

Reform Focus Groups (2002) www.doh.gov.uk/swqualification/focusgroups. pdf., accessed 27 Aug. 2005.

Reynolds, S. (2000) 'The Anatomy of evidence-based practice: principles and methods', in L. Trinder and S. Reynolds (eds), *Evidence-Based Practice: A Critical Approach.* Oxford: Blackwell Science. pp. 17–34.

Rice, G. (2003) *Report on the Social Work Workforce for Topss England.* Huddersfield: University of Huddersfield for Topss England.

Ritchie, J. (2003) 'The applications of qualitative methods to social research', in J. Ritchie and J. Lewis (eds), *Qualitative Social Research: A Guide for Social Science Students and Researchers.* London: Sage. pp. 24–46.

Roberts, H. (2004) 'Health and social care', in S. Fraser, J. Lewis, S. Ding, M. Kellett and C. Robinson (eds), *Doing Research with Children and Young People.* London: Sage in association with Open University. pp. 239–54.

Roe, W. (Chair) (2006) *Report of the 21st Century Social Work Review: Changing Lives,* Edinburgh: Scottish Executive.

Rogers, T. (2004) 'Managing in the interprofessional environment: a theory of action perspective', *Journal of Interprofessional Care,* 18(3): 239–49.

Rosen, M. (1991) 'Coming to terms with the field: understanding and doing organizational ethnography', *Journal of Management Studies,* 28(1): 1–24.

Sackett, D. L., Richardson, W. S., Rosenberg, W. and Haynes, R. B. (1977) *Evidence-Based Medicine: How to Practice and Teach EBM.* New York: Churchill Livingstone.

Save the Children (2004) *So You Want to Involve Children in Research? A Toolkit Supporting Children's Meaningful and Ethical Participation in Research Relating to Violence Against Children.* Stockholm: Save the Children Sweden.

Schein, E. (1984) 'Coming to a new awareness of organisational culture', *Sloan Management Review,* 25(2): 3–16.

Schon, D. I. (1971) *Beyond the Stable State.* New York: Random House.

Schutz, A. (1973) *The Phenomenology of the Social World.* Evanstown, IL: Northwestern University Press.

Schutz, A. (1978) 'Concept and theory formation in the social sciences', in J. Brynner and K. M. Stribley (eds), *Social Research: Principles and Procedures.* London: Longman.

SCHWG (2004) *Social Services and Workforce Survey 2003: Report No. 33.* Social Care and Health Workforce Group, accessed at www.lg-employers.gov.uk, 2 Aug. 2004.

SCIE (2004) *Learning Organisations: A Self-Assessment Resource Pack.* London: SCIE.

Seebohm Report (1968) *Report of the Committee on Local Authority and Allied Personal Services* (Cmd 3703). London: HMSO.

Shah, N. (1989) 'It's up to you sisters: black women and radical social work', in M. Langan and P. Lee (eds), *Radical Social Work Today.* London: Unwin Hyman. pp. 188–91.

Shardlow, S. (1989) 'Changing values in social work education: an introduction', in S. Shardlow (ed.), *The Values of Change in Social Work.* London: Routledge.

Shaw, I. (2003) 'Cutting edge issues in social work research', *British Journal of Social Work Research*, 33(1): 107–16.

Shaw, I. (2004) 'Evaluation for a learning organization', in N. Gould and M. Baldwin (eds), *Social Work, Critical Reflection and the Learning Organization.* Aldershot: Ashgate. pp. 117–28.

Shaw, I, Arksey, H. and Mullender, A. (2004) *ESRC Research, Social Work and Social Care*, Bristol: SCIE.

Sheldon, B. (1986) 'Social work effectiveness experiments: review and implications', *British Journal of Social Work*, 16(2): 223–42.

Sheldon, B. and Chivers, R. (2000) *Evidence-based Social Care: A Study of Prospects and Problems.* Lyme Regis: Russell House Publishing.

Sheldon, B., Chivers, R., Ellis, A., Moseley, A. and Tierney, S. (2005) 'A pre-post empirical study of the obstacles to, and opportunities for, evidence-based practice in social care', in A. Bilson (ed.), *Evidence-based Practice in Social Work.* London: Whiting and Birch. pp. 11–50.

Shepperd, M. (2004) *Appraising and Using Research in the Human Services: An Introduction for Social Work and Health Professionals.* London: Jessica Kingsley.

Silverman, D. (1993) *Interpreting Qualitative Data.* London: Sage.

Simpkin, M. (1979) *Trapped Within Welfare.* London: Macmillan.

Smith, J. (2004) *Fifth Report: Safeguarding Patients: Learning the Lessons of the Past – Proposals for the Future, Cmnd 6394.* London: HMSO.

Smith, L. T. (1999) *Decolonizing Methodologies: Research and Indigenous People.* London: Zed Books.

Smith, R., Monaghan, M. and Broad, B. (2002) 'Involving young people as co-researchers: facing up to the methodological issues', *Qualitative Social Work*, 1(2): 191–207.

Social Care and Health Workforce Group (2004) *Social Services Workforce Series 2003: report no.33.* www.lg-emplyers.gov.uk, accessed 2 Aug. 2004.

Solomos, J. (2003) *Race and Racism in Britain.* Basingstoke: Palgrave Macmillan.

Steele, R. (2003) *A Guide to Paying Members of the Public who are Actively Involved in Research.* Eastleigh: Involve.

Stein, M. (1997) *What Works in Leaving Care.* Ilford: Barnados.

Stone, J. A. M., Haas, B. A., Harmer-Beem, M. J. and Baker, D. L. (2004) 'Utilization of research methodology in designing and developing an interdisciplinary course in ethics', *Journal of Interprofessional Care,* 18(1): 57–62.

Stuart, R. (1985) '(Academic) Finding out in management: What has research to do with managers'. Association of Teachers in Management Focus Paper.

Taylor, C. and White, S. (2000) *Practising Reflexivity in Health and Welfare: Making Knowledge.* Buckingham: Open University Press.

THES (2006) 'HEFCE funding allocations', *Times Higher Education Supplement,* 1732: 8.

Thompson, N. (1997) *Anti-Discriminatory Practice.* Basingstoke: Macmillan.

Thompson, N. (2000) *Understanding Social Work: Preparing for Practice.* Basingstoke: Palgrave.

TOPSS (1999) *Modernising the Social Care Workforce.* Leeds: TOPSS.

TOPSS (2002) *The National Occupation Standards for Social Work.* Leeds: TOPSS.

Torkington, C., Lymbery, M., Millward, A., Murfin, M. and Richell, B. (2003) 'Shared practice learning: social work and district nurse students learning together', *Social Work Education,* 22(2): 165–75.

Trend, M. G. (1979) 'On the reconciliation of qualitative and quantitative analyses', in T. Cook and C. Reichardt (eds), *Qualitative and Quantitative Methods in Evaluation Research.* Beverly Hills: Sage.

Trevillion, S. and Bedford, L. (2003) 'Utopianism and pragmatism in interprofessional education', *Social Work Education,* 22(2): 215–27.

Trinder, L. (2000a) 'Introduction: the context of evidence-based practice', in L. Trinder and S. Reynolds (eds), *Evidence-Based Practice: A Critical Approach.* Oxford: Blackwell Science. pp. 10–16.

Trinder, L. (2000b) 'Evidence-based practice in social work and probation', in L. Trinder and S. Reynolds (eds), *Evidence-Based Practice: A Critical Appraisal.* Oxford: Blackwell Science. pp. 138–62.

Trinder, L. (2000c) 'A critical appraisal of evidence-based practice', in L. Trinder and S. Reynolds (eds), *Evidence-Based Practice: A Critical Appraisal.* Oxford: Blackwell Science. pp. 212–41.

Tuhiwai Smith, L. (1999) *Decolonizing Methodologies: Research and Indigenous Peoples.* London: Zed Books.

Underdown, A. (1998) *Strategies for Effective Offender Supervision: Report of HMIP What Works Project.* London: Home Office.

United Nations (1989) *Convention on the Rights of the Child.* New York: United Nations.

UPIAS (1976) *Fundamental Principles of Disability.* London: Union of Physically Impaired Against Segregation.

Walter, I., Nutley, S., Percy-Smith, J., McNeish, D. and Frost, S. (2004) *Knowledge Review 7: Improving the Use of Research in Social Care Practice.* London: SCIE.

Warwick, D. P. (1982) 'Tearoom trade: means and ends in social research', in M. Bulmer (ed.), *Social Research Ethics.* London: Macmillan. pp. 35–58.

Webb, S. (2001) 'Some considerations on the validity of evidence-based practice in social work', *British Journal of Social Work,* 31(1): 57–79.

Wilson, A. and Beresford, P. (2000) '"Anti-Oppressive practice": Emancipation or appropriation?', *British Journal of Social Work,* 30(5): 553–74.

Young, A. Hunt, R. and McLaughlin, H. (2006) 'Exploring models of D/deaf service user involvement in translating quality standards into local practice', *Social Work and Social Science Review*

Zwarenstein, M., Atkins, J., Barr, H., Hammick, M., Koppell, I. and Reeves, S. (1999) 'A systematic review of interprofessional education', *Journal of Interprofessional Care,* 13(4): 417–31.

Index

'abnormal science', 41
academic freedom, 68, 71
academic study of social work, 19–23
Ackroyd, S., 34
action research, 154, 172
Adams, R., 5
aggregate net-effect problem, 80
Alder Hey Hospital, 47
alienation, 117
Alzheimer's, 50
anonymity of research participants, 61
anti-discriminatory practice, 115, 129–30, 181–2
anti-oppressive practice, 51, 114–32, 181–2; concerns about, 120
anti-oppressive research, 121–3, 130–1
Antman, E., 73
Argyris, C., 172
Arnstein, S., 96
Association of Directors of Social Services, 93
asylum seekers, 119
auditing of services, 17
Australia, 22, 139, 179
Austria, 22
authorship of research, 68–9
autonomy, individual, 53

Bacon, Francis, 25
Baldwin, M., 172
Bangor, University of Wales at, 23
Banks, S., 50–4
Barclay, Peter, 2
Barn, R., 118, 122, 129–30
Barr, R., 140
Barratt, M., 173
barriers to incorporation of research into practice, 151–5, 183
  organisational issues, 164–6
  researcher issues, 161–4
basic research, 186
Bateson, G., 172
Becker, S., 10
Beresford, P., 98, 120, 127
Best Value process, 6, 17, 57, 100
biased research, 101, 122
Biestek, F., 49–50, 70
bio-medical research, 47
Boushel, M., 118
Bristol University, 23
British Association of Social Workers (BASW), 53, 55, 70, 122
British Journal of Social Work, 21
British Medical Association, 136
British Psychological Association, 47
British Sociological Association, 47
Bross, D. C., 82
'brute facts', 26, 29–30
Bryman, A., 10
'bugging', 59
Butler, I., 6–7, 47–8, 53–5, 60–1, 68, 70, 107, 122, 186
Butrym, Z., 4
Butt, J., 122

Cabot, R. C., 72–3
Cambridge Somerville Youth Study, 70
Cambridge University, 18
Campbell, D. T., 42
Campbell Collaboration, 74
casework principles, 49–50
Central Council for Education and Training in Social Work (CCETSW), 51, 115
Changing Lives report (2006), 185
Chapman, T., 76
child protection work, 82, 118
Children Act (1989), 4–6
children and young people's involvement in research, 105–11; advantages and disadvantages of, 106–9; as employees, 109; mandating of, 106–7

Children and Young People's Unit, 106
children's trusts, 8–9, 138
Chinese communities, 122
Chivers, R., 77, 152
Christensen, P., 95
Clark, A., 96
Clifford, D., 131
Cochrane Collaboration, 74, 85, 140
codes of practice, 47–55, 58–61, 115, 122
cognitive behavioural research, 76
Coleman, M., 120
collaboration in research, 98–9, 130; with young service users, 107, 109
Commission for Social Care Inspections, 6
'commonsensism', 31
Community Care (magazine), 22, 84, 153
Comte, Auguste, 25
confidentiality, 61–2, 69–79, 110
consultation, 97–8
control groups, 33
covert methods of research, 59–61
Crime and Disorder Act (1998), 134
critical appraisal of social work research, 155–61
Croft, S., 98, 127
Cronin, A., 38

Davidson, J. O. C., 60
D'Cruz, H., 1, 12, 43–4
deception on the part of researchers, 60
deductivism, 27
Deetz, S., 137
degree qualifications in social work, 52, 136
Delanty, G., 26, 138
Denmark, 22
Denzin, N., 130–1
Department of Health, 6, 56
Descartes, René, 25
deutero-learning, 172
Diploma in Social Work, 51
direct payments to disabled people, 100
directors of social services, 4, 8
disability issues, 31, 100
disciplinary status of social work, 23–4, 137, 182–3
discrimination, 47, 114–22
dissemination of research results *see* publication
divorce, impact of, 107
Dominelli, L., 13, 114, 118
'double blind' studies, 33
double-loop learning, 172

Easterby-Smith, M., 35
Economics and Social Research Council, 18–19
emancipatory research, 47
'embedded research' model, 169–70, 174–6
empirical form of knowledge, 26
employer support for research, 19, 168
employment of young service users, 109
empowerment, 54, 127–30
England, H., 49
Enlightenment philosophy, 53
epistemology, 26
ethical issues, 46–71, 73, 94, 122, 179–80; before research starts, 55–61; during research process, 61–6; after completion of data collection, 67–70
ethics committees, 55, 57, 61, 105
ethnic conflict, 117
ethnography, 37
Everitt, A., 12, 130
evidence-based practice, 72–86, 145, 152, 180, 185; definition of, 73; future prospects for, 85–6; lack of data on, 81–3; scientific basis of, 80–1
evidence-based probation, 75–7
evidence-based social work, 77
existentialism, 3
expectations of research outcomes, 162
exploitative research, 60

falsification, 27
Farmer, E., 118
feminism, different perspectives of, 124–5
feminist research practice, 124–7
Finklestein, V., 100
Finland, 22
Fisher, M., 20
Fiske, D. W., 42
focus groups, 37–8
Foster, P., 122–3
Framework for the Assessment of Children in Need and their Families, 73
France, 22
France, A., 110
funding of research, 18–23, 179
Furedi, F., 57
future prospects for social work research, 184–6

Garrett, P. M., 120
General Household Survey, 34
General Social Care Council (GSCC), 8, 55, 90, 115
Germany, 22
Getting Research into Practice (GRIP), 150

Gibbons, J., 118
Gibbons, M., 143–6
Giddens, A., 25
Gilbert, N., 40–1
Gillon, R., 54
Glasgow University, 23
Gospineth, C., 154
Gould, N., 174
Guena, A., 22

'halo effect', 70
Hammersley, M., 30–1, 122–3
Hanley, B., 96–8, 130
Harvey, L., 124
Health Act (1999), 138
Healy, K., 120
Heath, A., 32
Heron, J., 131
hierarchy of evidence, 74, 86
Higham, P., 135, 139
Higher Education Funding Council (HEFC), 18, 20, 23
Hodson, R., 173
Hoffman, C., 154
Homan, R., 47, 54, 59 60, 67–8
Hong Kong, 22
Hough, M., 76
Hudson, B., 139
Hughes, J., 10
Hughes, J. A., 34
Hugman, R., 49, 53, 135, 139–40
humanistic approach to research, 28
Humphreys, L., 47
Humphries, B., 60, 119–20, 184
Husband, C., 53
hypothetic-deductivism, 27

IMPACT study, 75
Imperial College, 18
incommensurability, 41, 45
indicative knowledge, 82
indigenous peoples, 117
individualisation, 51
informed consent, 58–63, 69, 109–10
inspection of social services, 6
institutional racism, 118
intellectual property rights, 56
interdisciplinary practice, 134–42
interdisciplinary research, 141–8, 182–3
International Association of Schools of Social Work (IASSW), 2–4, 178
International Federation of Social Workers (IFSW), 2–4, 178
Internet resources, 21–2, 155, 165–6
interpretivism, 28–31, 40–1, 45, 179
interview research, 39, 64
intranet systems, 22
intrusive methods of research, 60

Jackson, S., 32
Janzon, K., 165
Jay Committee (1979), 140
joint area reviews, 6
Joint University Council, 53
Jones, M., 1, 12, 43–4
Joseph Rowntree Foundation, 19

Kadushin, A., 39
Keele University, 23
Kimmel, A. J., 70
Kirby, P., 102, 107
knowledge: frameworks for understanding of, 7; *logical* and *empirical* forms of, 26; modes of production of, 143–6
Kuhn, Thomas, 40–1

Law, S., 165
Layder, D., 60
Le Court home, 100
'learning organisation' concept, 171–5, 183
learning processes, 172
Leverhulme Trust Foundation, 19
Lewis, J., 56–9, 64, 66
Lincoln, Y. S., 130–1
Listening, Hearing and Responding, 106
logical form of knowledge, 26

MacDonald, G., 34, 43, 74–5, 80–1
McDonald, P., 120
Macey, M., 116
McGuire, J., 73–6
MacIntyre, A., 53
Mackie, J. L., 53
McLaughlin, H., 97, 110
Making Research Count (MRC), 83, 86, 166, 185
managerialism, 5, 165, 173–4, 183
Marsh, P., 20
Martin, M., 60
Marxism, 51
May, T., 28, 36
Mayer, J. E., 89
meaningful involvement, 102–5
Merrington, S., 76
Mills, D., 20
minimum wage provisions, 109
mixed economy of care, 5

Mode 1 and *Mode 2* knowledge production, 143–6, 183
modernisation agenda, 5, 7, 20, 181
moral dimensions of social work, 53; *see also* ethical issues
moral panic, 119
Moxon, E., 116
multidisciplinarity, 147
Muluccio, A., 82

National Health Service (NHS), 55, 57, 138
National Health Service and Community Care Act (1990), 4–5
National Institute for Social Work, 89
National Occupational Standards for Social Work, 116
national service frameworks, 90
Netherlands, the, 23
Newman, T., 156
'normal science', 41
Northern Ireland, 8–9
Nuffield Foundation, 19

Oates, R. K., 82
objectivity, 94, 122
Oliver, M., 31
Oliver, P., 62–3, 69
O'Neill, A., 122
ontology, 26
organisational culture, 164–6, 183
'organisational excellence' model, 170–1
Orme, J., 127
'otherness', 130–1
Owen, M., 118
Oxford University, 18

Paley, J., 41, 43
paradigms, 30, 40–1, 44–5
Parker, R., 82, 164
participant observation, 36–7, 59–60
participation, 'ladder' of, 96
patriarchy, 125–6
Pawson, R., 7
payments to research participants, 103–4
Payne, G., 32–5, 126
Payne, J., 32–5, 126
Payne, M., 2, 124
Pedler, M., 172
peer review, 20–1, 145, 155, 179, 183–5
Performance Assessment Framework (PAF), 6
performance indicators, 165
philosophies of research, 25–45, 179
Platt, Denise, 6
police checks, 110
'political' aspects of research, 162–4
Popper, K., 27–8
positivism, 25, 29–33, 40–1, 45, 82, 125, 179–80
practical application of research, 150–76
'practitioner researcher' model, 167–9, 174–6, 183
Preston-Shoot, M., 177
probation, evidence-based, 75–7
Professional Social Work (magazine), 22
professional status, 135–6, 185–6
protection: of research participants, 63–6; of researchers, 66; of young service users, 110–11
Prout, A., 95
publication of research results, 67–8, 103, 131–2
publications on social work, 20–2
Punch, M., 60, 70

qualifications in social work, 8, 52, 136
qualitative research, 35–43, 57, 74, 159–61
Quality Assurance Agency for Higher Education, 53, 116
quantitative research, 35–43, 74, 161, 180
quasi-experimental studies, 81–2, 157–9
Queen's University, 23
questionnaires, use of, 35
questions, *open* and *closed*, 35, 39–40

racism and anti-racist practice, 115–23
radical social work, 51
randomised controlled trials (RCTs), 33–6, 80, 85–6, 180
'rational actor' model, 79
Raynor, P., 75–7
recording of research data, 62
reflexivity, 1, 3, 37
refutability of theory, 27
registration and regulation in social work, 8, 185
relativism, 30–1, 53
representativeness of service users involved in research, 92–3
research: definitions of, 10–11; reasons for engagement with, 16–18; seen as a business, 18–20, 178–9
Research Assessment Exercise (RAE), 17–18, 23, 183
'research-based practitioner' model, 166–9

Research Governance Framework (RGF) for Health and Social Care, 10, 55–8, 103–4, 168
Research in Practice (RiP) partnership, 83–6, 166, 173, 185
research literacy, 154–6, 161, 186
research mindedness 12, 14, 186
research tools, 31–3
research Web, 185
respect for persons, 50, 53

Reynolds, S., 74
Rice, G., 152
Richmond, Mary, 72
Ritchie, J., 41
Roberts, H., 106
Roe Commission (2006), 134, 185
Rosen, M., 11
Rowntree, Joseph, 72; *see also* Joseph Rowntree Foundation

Sackett, D. L., 73
Salford University, 23, 140–1
Save the Children, 105, 107
Schein, E., 164–5
Schon, D. L., 172
Schutz, A., 29
scientific theory and scientific method, 25–7, 41
Scotland, 6–8
scrutiny commissions, 57
Seebohm Report (1968), 184
service users' involvement in research, 88–112, 181; alternative views of, 95–6; benefits of, 90–2; by way of control, 99–102; levels of, 96–102; limits to knowledge production from, 105; meaningful forms of, 102–5; reasons for lack of, 92–5
sharing of research data, 67
Shaw, I., 19, 43, 169
Sheldon, B., 34, 72, 77, 85, 152–3, 166, 175
Shepperd, M., 156
Silverman, D., 35–6
single-loop learning, 172
Smith, D., 49, 53
Smith, Sir Graham, 75
Smith, Dame Janet, 136
Social Care Institute of Excellence (SCIE), 6–7, 83–5, 173, 185
social care sector, 8
social injustice, 13, 182
Social Research Association, 47
social service departments, 4–9 *passim*, 182
social work: definitions of, 2–3; nature of, 79–80
Social Work and Society (journal), 21
social work research, distinctiveness of, 12–14
'social worker' as a protected title, 136, 185
sociological analysis, 146
Sociological Research Online (journal), 21
Solomos, J., 118
Somalia, 106
sponsors of research, 56–7, 67–8
staff development, 166
Stanley, S., 76
star ratings, 6, 184–5
Strydom, P., 26, 138
Stuart, R., 153–4
subjectivity, 12
supervisors, role of, 69
surveys, 34–5
systematic reviews, 74–5

Tavistock Institute, 100
Taylor, C., 37
Thatcherism, 115
Thompson, N., 3–4, 120
Thorpe, R., 35
Timms, N., 89
training: of service users involved in research, 94, 105, 108, 112, 181; of social workers, 152, 166
transdisciplinary research, 143–8
Trend, M. G., 43
triangulation, 41, 43
Trinder, L., 80–3
True project, 94
truth, correspondence theory of, 26
Tuhiwai Smith, L., 119
Tuskegee syphilis study, 46

Underdown Report (1998), 75–6
Union of Physically Impaired Against Segregation (UPIAS), 100
United Nations Convention on the Rights of the Child, 106
United States, 23, 106, 179
University College, London, 18
university research on social work, 18, 23
user-controlled research, 99–102

Valuing People strategy, 90
Vico, Giambattista, 28
voyeurism in research, 64

Wales, 8–9
Walter, I., 166–7, 174, 183
Warwick, D. P., 47
Webb, S., 79–81
'what works' initiatives, 75–9
White, S., 37
Wilson, A., 120
Wiltshire and Swindon Users' Network, 100
withdrawal from research by participants, 62–3
worldviews, 40, 123

xenophobia, 115

Young, A., 166
young people *see* children and young people's involvement in research
youth offending teams (YOTs), 5–6, 78, 134

Zwarenstein, M., 140